Portable Circular Sawing Machine Techniques

Portable Circular Sawing Machine Techniques

• • • • •

Roger W. Cliffe

Sterling Publishing Co., Inc.   New York

Other Books by Roger Cliffe
Radial Arm Saw Techniques
Table Saw Techniques

DEDICATION
*To my aunts and uncles—
all fine role models*

Edited by Michael Cea

Library of Congress Cataloging-in-Publication Data

Cliffe, Roger W.
    Portable circular sawing machine techniques / Roger W. Cliffe.
      p. cm.
    Includes index.
    ISBN 0-8069-6552-5 (pbk.)
    1. Circular saws. 2. Woodwork. I. Title.
TT185.C62   1988
684'.083—dc 19   87-33673
                                                                                       CIP

1   3   5   7   9   10   8   6   4   2

Copyright © 1988 by Roger W. Cliffe
Published by Sterling Publishing Co., Inc.
Two Park Avenue, New York, N.Y. 10016
Distributed in Canada by Oak Tree Press Ltd.
% Canadian Manda Group, P.O. Box 920, Station U
Toronto, Ontario, Canada M8Z 5P9
Distributed in the United Kingdom by Blandford Press
Link House, West Street, Poole, Dorset BH15 1LL, England
Distributed in Australia by Capricorn Ltd.
P.O. Box 665, Lane Cove, NSW 2066
Manufactured in the United States of America
All rights reserved

Contents

Preface 8

FUNDAMENTALS 9

1. History and Development of the Portable Circular Saw 10

2. Safety Guidelines 20
Factors Contributing to Accidents 20; Electrical Safety Procedures 21; General Working Environment 22; General Safety Procedures 24; OSHA Regulations 25

3. Circular-Saw Blades 27
Circular-Saw Terms 27; Common Blade Types 30; Selecting Blades 40; Blade Maintenance 41

HAND-HELD CIRCULAR SAWS 45

4. Introduction to the Portable Circular Saw 46
Saw Size 46; Types of Power 48; Classifications 50; Accessories 56

5. Basic Operations 69
General Safety Guidelines 69; Crosscutting 70; Ripping 75; Mitre Cuts 82; Cutting a Taper 87; Compound-Mitring 87; Cutting Plywood and Sheet Stock 90; Cutting Hardwood Plywood and Laminated Sheet Stock 97; Getting a Straight Cut 98

6. Advanced and Specialized Operations 100
Plunge-cutting 100; Cutting Irregular Shapes 102; Cutting Arcs or Circles 105; Cutting Joinery 105; Cutting Notches in Framing Members 116; Trimming Doors 120; Cutting Nonwooden Materials 127; Stationary Saw Guides and Devices 133

7. General Care and Maintenance of Your Portable Circular Saw   *154*
Changing the Blade   155; Adjusting the Saw   157; Lubrication of the Saw   158; Electrical Maintenance   159; Buying a Portable Circular Saw   160

MITRING MACHINES   *163*

8. Plunging or Chop-stroke Mitring Machines   *164*
Classifications   164; Parts   167; Accessories and Features   168; General Safety Guidelines   173; Common Operations   176; Cutting Crown Moulding   189; Making a Cope Joint   189; Brick-Mould Corners   194; Mitre-Box Picture Frames   194; General Trim Installation   194; Determining Mitre Cuts for Odd Angles   202; Cutting Metal and Plastic   204; Accessory Tables for Mitre Boxes   204; Maintenance of Chop-Stroke Mitring Machines   209; Purchasing a Chop-Stroke Mitring Machine   214; Specialty Mitre Machines   215

9. Pull-Stroke Mitring Machines   *217*
Parts   219; General Safety Guidelines   221; Common Operations   224; Maintenance of Pull-Stroke Mitring Machines   243; Accessory Tables   252

SPECIALTY AND STATIONARY CIRCULAR SAWS   *255*

10. Specialty and Stationary Circular Saws   *256*
Special-Purpose Saws   256; Machines of Unique Design   259; Portable Stationary Saws   265

PROJECTS   *277*

11. Accessories for Your Portable Circular Saws   *278*
Tips for Building Projects   278; Projects   278

APPENDICES *333*

Metric Equivalency Chart *335*

Glossary *336*

Index *344*

Photo Credits *351*

Acknowledgments *352*

Preface

With over 40 million portable circular saws and untold numbers of motorized mitre boxes, combination saws, and compound-cut saws available in the United States, one can see a need for *Portable Circular Sawing Machine Techniques*. Portable circular saws can be found in homes, in industrial settings and cabinet shops, and on construction sites. They are used by both experienced and inexperienced operators. *Portable Circular Sawing Machine Techniques* addresses the need for basic information and explains the cutting techniques that all portable circular-saw operators should master.

Included in the basic information core are topics such as:
1. Blade selection for portable circular saws.
2. Blade maintenance for portable circular saws.
3. Setup and maintenance of portable circular-sawing machines.
4. How to buy portable circular-sawing machines.
5. Accessories that can be used with portable circular-sawing machines.
6. Techniques for operating portable circular-sawing machines efficiently and correctly.
7. Safety tips and procedures for operating portable circular-sawing machines.

Included in the operations core are topics such as:
1. Ripping and crosscutting procedures.
2. Simple- and compound-mitring with the portable circular cutoff saw.
3. Using cutting guides and specialty accessories with the portable circular saw.
4. Cutting glass, slate, and steel with a portable circular saw.
5. Cutting notches, pockets, and joinery with a portable circular saw.
6. Crosscutting, mitring, and compound-mitring with the motorized mitre box.
7. Using stands and accessories with the motorized mitre box.
8. Crosscutting, mitring, and compound-mitring with compound-cut saws.
9. Using hand tools to cut cope joints to match installed trim.

Portable Circular Sawing Machine Techniques was written for both the novice and experienced woodworker. The collection of special set-ups and techniques represents the knowledge of many experienced woodworkers and carpenters with whom I have had the pleasure of working. While many seasoned woodworkers will be familiar with some of these operations, it is certain they will not be familiar with all of them. Also, many operations that are only performed occasionally are easy to forget; this book will serve as a thorough guide to those operations.

Fundamentals

1
History and Development of the Portable Circular Saw

In 1924, the first portable circular saw was marketed by the Michel Electric Handsaw Company, which was cofounded by Edmond Michel, the inventor of the portable circular saw, and Joseph Sullivan, a land developer.

Two years after the saw was marketed, Edmond Michel left the company to work on other inventions. At that time, the Sullivan family renamed the company Skilsaw™. Many people still refer to any portable circular saw as a Skilsaw™. This term, however, is not a generic term; it is a trade term used by the Skil Corporation.

As time went on, the Skil Corporation perfected its design. In 1937, Skil introduced the famous Model 77 worm-drive portable circular saw. This saw is still in production today with few modifications (Illus. 1).

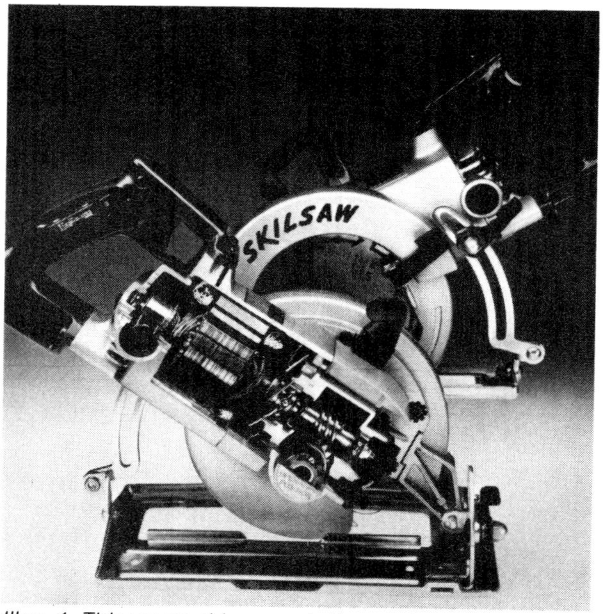

At the same time the Michel Electric Handsaw Company was developing the first portable electric circular saw, Bill Casey was designing a similar device. His tool is often referred to as the Casey Electric Hand Tool (Illus. 2). The machine was patented in 1919.

The Casey machine is a specialty portable circular saw. It was designed to make straight line cuts for openings in doors, but it also cuts dadoes and V grooves (Illus. 3). Several guide tracks are available for use with the Casey Electric Hand Tool. They control the path of the tool and increase the accuracy of the dado or cut. The Casey Manufacturing Company is still in business in Oshkosh, Wisconsin, selling and leasing its machines.

Porter Cable also contributed greatly to the advancement of portable circular saws. In 1929, it developed the anti-kick clutch for circular saw safety (Illus. 4), which reduced the chance of a kickback when the operator used a portable circular saw. In

Illus. 1. This worm-drive saw is the contemporary model of the original Model 77 worm-drive saw.

Illus. 2. This Casey Electric Hand Tool, which was patented in 1919, is a specialty tool designed to be controlled by a guide.

Illus. 3. Note the dado cutter used on the Casey Electric Hand Tool. Dado cutters are rarely used on typical portable circular saws.

Illus. 4. The spring-loaded arbor washer minimizes the chance of a kickback. This anti-kickback feature was developed by Porter Cable in 1929.

Illus. 5. This microprocessor-controlled portable circular saw was developed by Porter Cable in 1983.

1938, Porter Cable developed and marketed the first helical-gear portable circular saw with a right-hand drive. Until this saw was marketed, all saws used the worm drive. To date, all portable circular saws are either worm-gear or helical-gear drive.

Porter Cable's latest contribution was in 1983, when it marketed the first microprocessor-controlled portable circular saw (Illus. 5). The microprocessor allows the saw to start slowly and gradually increase speed. It also senses cutting resistance and increases or decreases cutting speed accordingly. An indicator light tells the operator whether or not the saw is working under ideal conditions.

In 1962, Skil Corporation marketed the first line of double-insulated professional tools. Double-insulated portable circular saws do not require a grounded outlet. The saw itself is protected by a nonconducting case. This design makes electric shock almost impossible. Skil has continued to market well-engineered portable circular saws (Illus. 6).

Another advancement that makes electric shock impossible is the cordless or battery-operated saw (Illus. 7). During the late 1970s and early 1980s, the technology related to batteries advanced rapidly. This led to the development of a small, light, and powerful battery that was capable of driving a portable circular saw (Illus. 8).

Motors also got smaller, but more powerful, during the same time period. This led to the development of portable circular saws with blades 10 inches in diameter and over (Illus. 9 and 10). Electric brakes were added to make these large saws safer. The electric brake stops the blade as soon as the trigger switch is released.

Illus. 6. Skil has marketed well-engineered portable circular saws since 1924.

Illus. 7. Cordless or battery-operated portable circular saws make electric shock impossible. Great advancements have recently been made in the technology used to make cordless saws. These saws are small and quite powerful.

Illus. 8. The battery used in this saw is light and powerful. This saw is capable of trimming and light framing jobs.

Illus. 9 (above left). This portable circular saw turns a 25¼-inch saw blade. Illus. 10 (above right). This portable circular saw turns a 16-inch blade. It is equipped with an electric braking system.

Portable circular saws have also gotten smaller. Many saws have blades as small as 4 inches in diameter. Today, there are many specialty saws for such uses as panelling (Illus. 11). These saws are smaller and easier to handle. The operator feels less fatigue when he uses them.

Illus. 11. Portable circular saws vary in size, blade diameter, and the materials used for manufacture. Advanced technology has made it possible to use plastic for saw housings and guards. Plastic has also made the saw shockproof.

Today, many manufacturers produce portable circular saws (Illus. 12–17). The saws vary in the ways they are built, their horsepower, and in the materials used to make the base and housing. New blades have also been developed, including blades that cut curves (Illus. 18), diamond-cutting blades, and blades wrapped with chain-saw teeth (Illus. 19).

As the design of the portable circular saw was perfected, designers used the saw on other devices. Most of these devices were developed from 1960 to the present. One of the most popular adaptations was the portable circular saw's use on a mitre box (Illus. 20 and 21); this combination is known as a motorized mitre box.

The motorized mitre box increased the quality of the mitre cut and reduced the amount of time required to cut a mitre. The first motorized mitre boxes started using 9–10-inch blades, but today some machines use blades as large as 16 inches in diameter. These machines utilize an electric brake which stops the blade as soon as the trigger switch is released.

Compound-mitre cutting saws became the next advancement after the motorized-mitre box. A compound mitre is a mitre cut at two angles. Compound mitres are cut on stock that is turned up at an angle. Simple mitres lie flat.

Illus. 12. This saw uses a 6½-inch saw blade. It has a solid metal base or shoe.

Illus. 13 (above left). This 7¼-inch portable circular saw has a metal guard and a plastic motor housing. It is a contractor-grade saw. Illus. 14 (above right). This 7¼-inch portable circular saw is also a contractor-grade saw. It has a drop-shoe design. Design features are discussed in Chapter 4.

Illus. 15. This contractor-grade portable circular saw has a pivot-foot design. Saw design or features will determine selection and use of the portable circular saw.

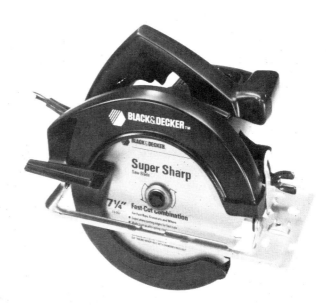

Illus. 16. This 7¼-inch portable circular saw is considered a homeowner's saw. It has a plastic housing and guard and a pivot-foot design.

Illus. 17. This pivot-foot saw turns a 7¼-inch blade. It is also a common-sized portable circular saw.

Illus. 18 (above left). The triangular shape of this blade enables it to cut an arc or irregular curve. Some of these blades now have carbide tips. More information about saw blades can be found in Chapter 3. Illus. 19 (above right). The saw blade shown here has chain-saw-type teeth wrapped around it. It cuts wet wood without binding.

Illus. 20. This motorized mitre box is an adaptation of the portable circular saw. It controls the blade's path to make quality mitre cuts.

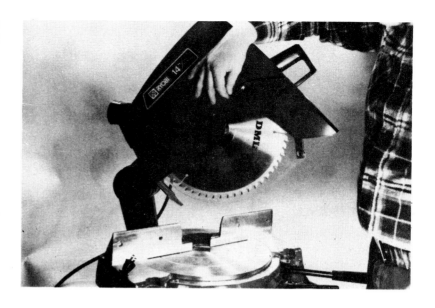

Illus. 21. This motorized mitre box uses a larger blade and cuts thicker stock than the one in Illus. 20.

Some compound-mitre saws closely resemble the motorized mitre box (Illus. 22), while others look like a miniature radial-arm saw (Illus. 23). Other compound-mitre saws have an innovative design that allows for compound-mitre cuts on a much wider piece of stock (Illus. 24).

All mitring machines can be categorized as chop-stroke or pull-stroke machines. On chop-stroke machines, the blade comes down on the work; on pull-stroke machines, the blade cuts across the work. Pull-stroke machines are usually compound-mitring machines. Chop-stroke machines may be either simple- or compound-mitring machines. All of these machines are designed with portability in mind.

Another modification of the portable circular saw is the Crain Model 800 Super Saw (Illus. 25). This saw is used by workers in the floor-covering industry. It notches baseboard (Illus. 26), and trims doors and door jambs (Illus. 27–29). The modifications make this tool an excellent addition to the portable circular-saw family. The typical portable circular saw would be unable to perform any of the functions just described.

The modified portable circular saw shown in Illus. 30 is designed to shape or machine notches in rafters and other framing stock. A cutterhead replaces the saw blade. This cutterhead is similar to the one that is on a jointer or shaper.

Many stationary machines have also been designed for portability. The standard 10-inch table and radial-arm saws of the past are often replaced with a portable machine with an 8-inch blade (Illus. 31 and 32). The improvements in electric motors allow a much smaller motor to produce more power. Aluminum and other materials replace the cast iron once found in these machines, with the result that these machines have found a new market.

These smaller table and radial-arm saws have been widely accepted by the woodworker with limited shop

Illus. 22. This motorized mitre box can cut compound mitres. The blade tilts and the table turns. This type of mitring machine is quite new to the market.

Illus. 23. This compound-cut saw will also cut compound mitres. It is designed like a portable miniature radial arm saw.

Illus. 24. The Sawbuck™ is designed to cut compound mitres on wide pieces. This trim and framing saw is portable and easy to set up.

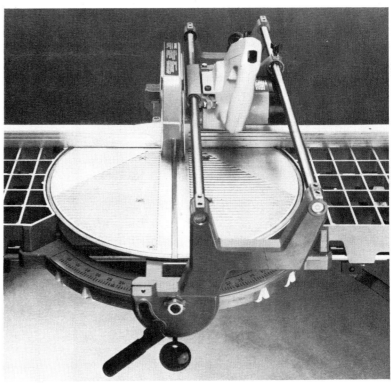

Illus. 25. This modified portable circular saw is used by people in the floor-covering industry for specialty cuts. It is known as the Super Saw.

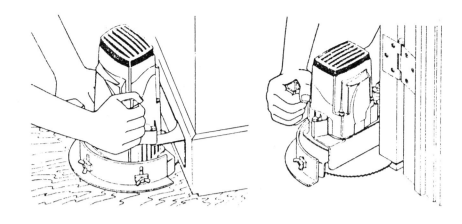

Illus. 26 (right). The Super Saw can be used to notch baseboard for underlayment or for other floor treatments. Illus. 27 (far right). The Super Saw can trim a door while it remains on the hinges. You cannot trim the wrong end of the door this way.

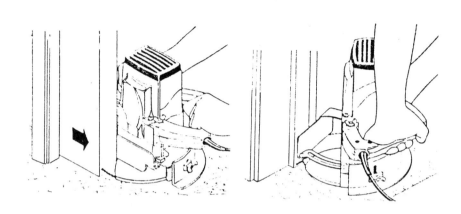

Illus. 28 (right). A pocket door can also be trimmed while it remains on its track. This is the only portable circular saw that can perform this job. Illus. 29 (far right). Door jambs and moulding can also be trimmed without difficulty. This allows new flooring to slide under the jamb, something you'll find very helpful when you're remodelling.

Illus. 30. This portable circular saw has a cutterhead attached instead of a blade. It is used to notch rafters.

Illus. 31. This portable 8¼-inch table saw will do most of the jobs that a 10-inch stationary machine will do.

Illus. 32. This 8¼-inch radial arm saw is a bench-top portable machine. It, too, will do most of the jobs done by a 10-inch stationary machine.

space. These machines can be set up when needed, but stored easily when not in use. Some have accessory stands which allow them to collapse into a unit that can be conveniently stored.

At the time this book was written, there were approximately 40 million portable circular saws, 10 million table saws, and 7 million radial-arm saws in the United States. There are numerous accessories available for all portable circular saws, motorized mitre boxes, compound-mitring machines, and portable table and radial-arm saws. In this book I present as many of these accessories as space permits. I also present techniques for safe and efficient use of all portable circular saws, motorized mitre boxes, and compound-mitre saws.

2
Safety Guidelines

Factors Contributing to Accidents

Portable circular-saw accidents can occur with either a novice or experienced operator. The novice's accident is usually caused by a lack of knowledge or experience. He may have little knowledge of portable circular saw safety. However, a knowledge of safety procedures helps the novice identify and avoid an accident-producing situation.

The experienced operator's accident is usually caused by carelessness or an outright violation of the safety rules. When an experienced operator attempts and gets away with safety-rule violations, they soon become common practice. This is when an accident is likely to occur.

All accidents have other contributing factors, among which are the following:

1. *Working While Tired or Taking Medication.* Whenever you are tired, stop or take a break. Accidents are most likely to happen when you are tired. Medication or alcohol can affect your perception and reaction time. They can also affect your balance and judgement, both of which are necessary for safe operation of any woodworking tool or machine.

2. *Rushing the Job.* Trying to finish a job in a hurry leads to errors and accidents (Illus. 33). The stress of rushing the job also leads to early fatigue, which can also lead to accidents. Before rushing a job, ask yourself how quickly you can complete the job if you are injured or if the work is ruined.

3. *Inattention to the Job.* Daydreaming or thinking about another job while operating a circular saw can contribute to accidents (Illus. 34). Daydreaming frequently occurs when you make repetitive cuts or perform production operations. Be extremely careful while making them; maintain your concentration. When this becomes difficult, take a break.

Illus. 34. Repetitive cuts become automatic after a short while. This leads to daydreaming, and possibly an accident. Maintain your concentration when doing repetitive work.

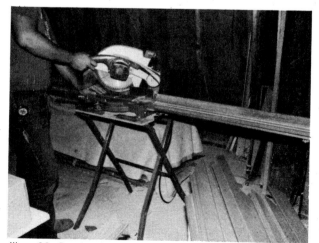
Illus. 33. Production jobs are generally conducted at a brisk pace. This can lead to fatigue, errors, and accidents.

4. *Distractions.* You can be distracted in the shop by unfamiliar noises, by doors opening or closing, or when you have a conversation with others. You can also be distracted by loud noises, deliveries, the movement of equipment. Be sure to shut off the saw before you converse with someone or investigate an unfamiliar noise. Personal experience has proven that

babysitting and woodworking (however small the project) do not mix. They are both full-time jobs.

5. *General Housekeeping.* A dirty or cluttered work area provides tripping and balance hazards. An accumulation of dust can become a breathing or vision hazard. Keep the shop or work area neat and clean. It is far more pleasant and safer to work in a clean area. When working on site, set up the circular saw on a level surface. Clean up the area frequently. Small cut-offs pile up quickly and are balance and tripping hazards. Keeping the area neat will eliminate tripping hazards.

Electrical Safety Procedures

The fact that circular saws are electrically powered means that you, the operator, could be subject to electrical shock. Work in the field is seldom done under ideal conditions, but it should be done with complete adherence to electrical safety procedures.

In most cases, portable equipment is plugged into extension cords for use in the field. When using an extension cord in the field, observe the following rules:
1. Make sure that the cord is intended for outdoor use.
2. Make sure that the cord is of the proper size and wire gauge so that it can be operated safely and efficiently. Cords of improper size will reduce the saw's power and may cause the tool to burn out prematurely. See Table 1 to determine the correct-size extension cord for the saw you are using.
3. Use cords that are approved or "listed" by groups such as Underwriters Laboratories or the Canadian Safety Association.
4. Inspect cords before and after each use. Make sure that there is no damage to the plug, outlet, or insulation. Replace damaged or defective cords immediately (Illus. 35). Destroy the defective cord so that it is not used by someone else.

Make sure that all circuits are grounded correctly before plugging a saw or extension cord into the circuit. Some saws are double-insulated and do not require grounding. These tools have a symbol (a square within a square) that indicates double insulation. Double-insulated tools have a nonconducting case that encloses all electrical parts. This eliminates the chance of contact with the electric current.

Some circuits do not have a ground and may require an adaptor or "pigtail." Make sure that the ground wire off the adaptor is properly grounded. When in doubt, have the electrical system inspected by a licensed electrician. The use of adaptors is dwindling because of grounded electrical systems, and is not permitted in Canada.

In many field operations, Ground Fault Circuit Interrupters (GFCI) are used to reduce the electrical hazard. The GFCI measures the amount of current going to the tool and returning to the GFCI. When there is an imbalance in the current going and returning, a ground fault exists. A ground fault interrupts the current going through the GFCI. The GFCI must be reset manually after an interruption.

When a ground fault occurs, the GFCI assumes a

Important Information about the Use of Extension Cords

Use the right extension cord. An extension cord should have a suitable wire size for the overall cord length and tool amperage rating. This is to prevent a serious voltage drop, power loss and possible motor damage. Generally, heavier gauge wire is required as cord length increases. Use the recommendations in this table. This table is based on limiting line voltage drop to 5 volts at 150% of rated amperes.

Table 1. Use this chart to select the correct size extension cord to be used with your portable circular saw.

Extension Cord Length, ft.	Amperage Rating of Tool					
	0–2	2.1–3.4	3.5–5	5.1–7	7.1–12	12.1–16
	Recommended Wire Size (Gauge)					
25	18	18	18	18	16	14
50	18	18	18	16	14	12
75	18	18	16	14	12	10
100	18	16	14	12	10	8
150	16	14	12	10	8	8
200	16	14	12	10	8	8
300	14	12	10	8	6	4
400	12	10	8	6	4	4
500	12	10	8	6	4	2
600	10	8	6	4	2	2
800	10	8	6	4	2	1

■ Not normally available as flexible extension cord.

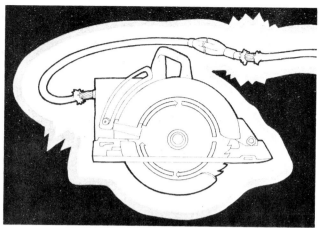

Regularly check tool power cords and extension cords. Never use a tool if there is any sign of cord damage. Worn and damaged cords must be replaced.

Illus. 35. Replace a defective or damaged cord immediately. Defective cords lead to electric shocks.

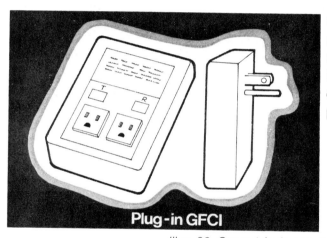

Ground Fault Circuit Interruptors provide another means to reduce serious electrical shock while using a power tool. They are easily installed between the power receptacle and cord. Leakage of current will shut off the tool, preventing possible serious injury.

Illus. 36. Ground fault circuit interrupters are required by OSHA at all construction sites.

short or ground through another source (such as the operator). This stops the current and stops the shock almost instantly. Ground fault circuit interrupters are required by OSHA (Occupational Safety and Health Act) at all construction sites (Illus. 36).

The environment around an electrical current can also promote possible shock or electrocution. Rain, dampness, or wet conditions encourage the conduction of electricity. Using an electric saw on damp or wet earth or using damp tools can also conduct electricity away from the motor to the operator. When conditions in the environment favor electrical shock, consider using a cordless, internal-combustion, or pneumatic, saw. These tools will not cause electrical shock.

General Working Environment

The environment you work in is one factor in the safe operation of portable circular saws. The work should be at a comfortable height, and should be stable (should not rock or shift). Some woodworkers prefer the work at about knee height, while others prefer it at about waist height. Larger pieces are usually easier to work on if they are below waist level because your reach is greater when the work is a little lower.

Clamps and other holding devices should be readily available near the work area. With these devices, you can secure awkward parts or small parts so that your hand or some other extremity is not placed too close to the saw blade. When working on an awkward part, you may not be too close to the work, but will still be unable to hold it because it's uncomfortable. Clamps eliminate this problem.

Mitre saws and compound-mitre saws should be positioned for comfort (Illus. 37); 30 inches is generally a good height. Some machines can be slightly higher or lower. Mitre saws should not be used on the floor or ground; this is an awkward work position, and one hardly looks like a skilled woodworker when working on the floor.

Mitre saws should be firmly anchored to their stand, and the stand should be positioned so that it does not rock. Long parts should be supported with an auxiliary table or a dead man. These supports will increase control over the work and reduce the chance of a mishap.

When setting up a work station, consider traffic and material flow. Moving material should not present a hazard to other workers, and the material should be handled as little as possible. Handling material twice to cut it once is a waste of labor and increases the chance of a mishap.

The work station should have provisions for scrap. A container for cutoffs should be near it, and should be emptied regularly. When the work area is clean, work progresses normally with less chance of a trip or mishap (Illus. 38).

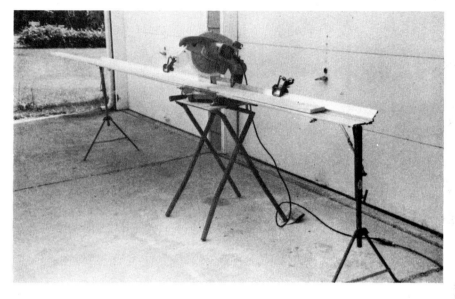

Illus. 37. Position your saw for comfort. Working on the ground can not only cause accidents, it also looks unprofessional. It can also lead to early fatigue.

Illus. 38. Make provisions for scrap and empty those containers regularly. This reduces the chances of tripping or mishaps.

General Safety Procedures

When working with portable circular saws, there are some general rules that you should observe. These rules apply to all portable circular saws. Safety information related to a specific tool will be found under the discussion of that tool in a later chapter.

1. *Know your tool.* Read and understand the owner's manual and any warnings or cautions on the tool before you operate it. Practice on scrap until you feel comfortable with the tool.

2. *Make all adjustments with the power disconnected.* When changing blades, adding accessories, or making adjustments, disconnect the saw. When the saw is plugged in, there is always a chance of it accidentally starting. When working with cordless saws, remove the battery.

3. *Make sure that the accessory is compatible.* When installing a blade or accessory on your saw, make sure that it was designed for use on that tool. Check the rated speed (rpm's) to be sure it is compatible with the saw speed (Illus. 39). Check your owner's manual to be sure the accessory is recommended.

4. *Maintaining Your Saw.* Make sure that the switch, guard, and brake work properly. If the saw has a clear plastic guard, clean it periodically with mild detergent. Never use a saw that is not in good working order. Keep a clean, sharp blade on the saw. Any pitch accumulation on the blade should be removed immediately. Check the arbor nut or bolt to be sure it is tight. Power saws with an electronic brake tend to loosen the arbor nut. When your saw is not in use, store it in a clean, dry place. Inspect the saw carefully before using it again.

5. *Dress for the Job.* When working with portable circular saws, remove loose clothing, jewelry, watches, and gloves. Button long sleeves or roll them up past your elbows. Long hair should be restrained. Loose objects can get caught in the moving saw blade, and pull you in. Wear safety shoes or appropriate footwear; this minimizes the chance of injury and improves your balance. When working with treated lumber, be sure to wash your clothing immediately after working. Wash carefully before eating, drinking, or smoking.

6. *Protect Yourself.* Always wear protective glasses when working with portable electric saws. If the area is noisy, wear ear plugs or muffs to preserve your hearing and minimize fatigue. Gloves are all right for handling rough lumber, but they should never be worn when sawing. When working with fine-cutting blades or modified woods such as particle board or treated lumber, wear a dust mask or respirator. This will protect your lungs, nose, and throat from irritation or damage (Illus. 40).

Illus. 39. Make sure blades or accessories are compatible with the saw you are using. All blades list maximum safe speed (rpm's); make sure that your blade is above the saw speed.

Illus. 40. Dust masks or respirators can be used to protect your lungs, nose, and throat from harmful or very fine dusts.

7. *Maintain Your Balance.* If your footing is unsecured when you are working, you can trip or fall. Overreaching invites a kickback or other mishap because you lose a great deal of your strength and balance.

8. *Inspect Your Stock.* Before you saw any stock, look it over. Loose knots, twists, cupping, and rough or wet lumber can mean trouble. Loose knots can be ejected by the saw. They can also shatter carbide tips if they move during the cut. Rough warped or wet lumber can throw the saw back at you during the cut (Illus. 41). Small pieces can also mean trouble. Machining them puts your hands too close to the blade.

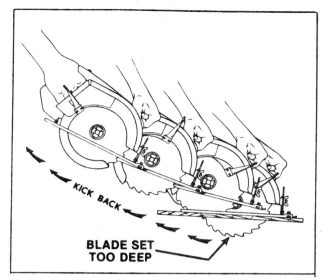

Illus. 41. A kickback can be caused by rough, twisted, or wet lumber. Inspect your stock carefully before cutting, and keep your blade exposure to a minimum.

It is much safer to cut a small part off a larger workpiece.

9. *Use Control Devices.* Devices like straightedges, clamps, and push sticks make sawing safer and easier (Illus. 42). A push stick can hold the work against the table and fence of a mitre saw, and a straightedge can guide a portable circular saw along a straight line. Control devices can keep your hands well away from the cutting area, and provide positive control.

10. *Keep a Safe Distance from the Cutting Zone.* Keep your hands 8 inches or more from the blade at all times; this allows a margin for error. When your hands are a safe distance from the blade, there is always time to react to a hazardous situation. Remember that it is difficult to cut yourself when both hands are on the portable circular saw. When the saw has two handles, use them (Illus. 43)! Clamp stock to keep your hands clear of the cutting zone.

11. *Never Force the Saw.* Always let the blade do the cutting. Forcing the saw invites a mishap and causes undue wear on the saw and blade.

12. *Think About the Job.* When performing a new operation, think about the job before you begin. Ask yourself, "What could happen when I . . . ?" Questions like this will help you identify and avoid an accident-producing situation. If you have a premonition of trouble, stop! Avoid any job that gives you a bad feeling. Try setting up the job another way, or ask some other experienced operator for an opinion (Illus. 44).

OSHA Regulations

Portable circular saws used in industry and on construction sites are covered by OSHA (the Occupational Safety and Health Act). This act regulates the setup and use of industrial equipment, and is enforced by the Occupational Safety and Health Administration, which is part of the United States Government. OSHA lists specific requirements for portable circular saws. In addition, there are some general requirements for all woodworking shops.

Some of the requirements listed by OSHA for portable circular saw use are as follows:

1. *Branch circuits. Ground-fault protection for personnel on construction sites.* The employer shall use either ground-fault circuit interrupters or an assured equipment grounding-conductor program to protect employees on construction sites. These requirements are in addition to any other requirements for equipment grounding conductors.

2. *General requirements.* Each employer shall be responsible for the safe condition of tools and equipment used by employees, including tools and equipment which may be furnished by employees.

3. *Portable powered tools (portable circular saws).* All portable, power-driven circular saws having a blade diameter greater than two inches shall be equipped with guards above and below the base plate or shoe. The upper guard shall cover the saw to the depth of the teeth, except for the minimum arc required to permit the base to be tilted for bevel cuts. The lower guard shall cover the saw to the depth of the teeth, except for the minimum arc required to allow proper retraction and contact with the work.

Illus. 42. Control devices make sawing safer and easier. There is also increased accuracy when control devices are used. Be sure to keep your hands at least 8 inches away from the blade and out of the kickback zone.

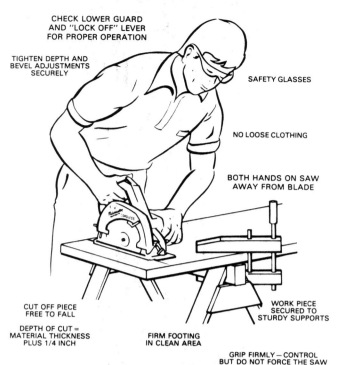

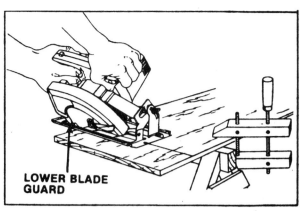

Illus. 43A and B. When you keep both hands on the saw, it is difficult to cut them. Use both handles whenever possible.

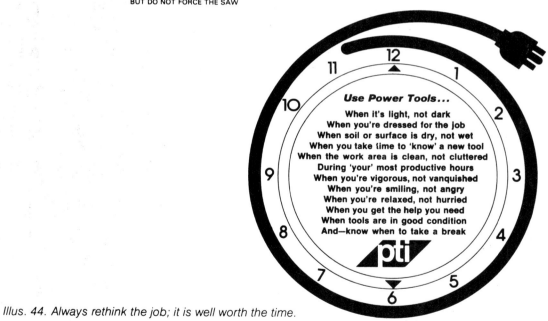

Illus. 44. Always rethink the job; it is well worth the time.

When the tool is withdrawn from the work, the lower guard shall automatically and instantly return to covering position.

4. *Switches and controls.* All hand-held powered circular saws having a blade diameter greater than two inches shall be equipped with a constant-pressure switch or control that will shut off the power when the pressure is released.

5. *Operating control.* The operating control on hand-held power tools shall be so located as to minimize the possibility of its accidental operation, if such accidental operation would constitute a hazard to employees.

6. *Cracked saw blades.* All cracked saw blades shall be removed from service.

3

Circular-Saw Blades

When you work with wood, the type of cut usually determines what type of blade you will select. These blades will differ according to the type of work and the type of portable sawing machine being used. To help understand what type of blade must be selected, it is important to know a little about the nature of the material you will be cutting.

Wood is a stringy material. If you break a piece of it, the stringy fibres make it difficult to get a clean break. If you split it, the split is clean because the fibres run parallel to the separation. When crosscutting wood (cutting it across the grain or fibres), use a crosscut blade (Illus. 45). The teeth are designed for crosscutting. Rip blades (for cutting with the grain or fibres) are designed for ripping only (Illus. 46), and do not crosscut efficiently. There are other specialty blades and combination blades designed to rip or crosscut. They are discussed later in this chapter. Remember, whatever type of sawing you are doing or

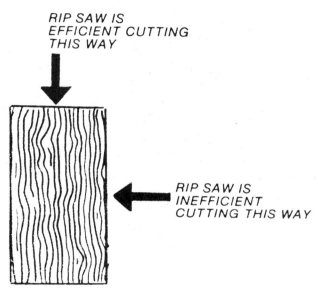

Illus. 46. When cutting with the stringy fibres, use a rip blade. A rip blade has chisel-shaped teeth.

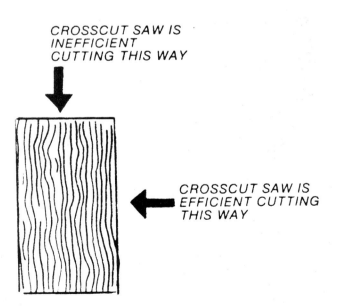

Illus. 45. A crosscut blade with pointed teeth is used to cut across the grain of stringy fibres.

whatever machine you are using, the saw blade is the most important link in the sawing process. To do the job, blades must have the correct tooth configuration. They must also be sharp and true.

Circular-Saw Terms

The cut made by a circular-saw blade is called the *kerf* (Illus. 47). The kerf must be slightly larger than the saw-blade thickness. The *tooth set* or *offset* is the bend in the tooth (Illus. 48). This set allows the blade to cut a kerf that is larger than the blade's thickness.

The teeth on a circular-saw blade are set in alternate directions. The set on blades used for portable circular saws is usually greater than the set on blades for stationary machines or portable mitring machines. This helps minimize the chance of a kickback.

The *gullet* is the area behind the cutting edge of the tooth (Illus. 49). It carries away the sawdust cut by the tooth. The larger the tooth, the larger the gullet.

The *hook angle* is the angle of the tooth's cutting

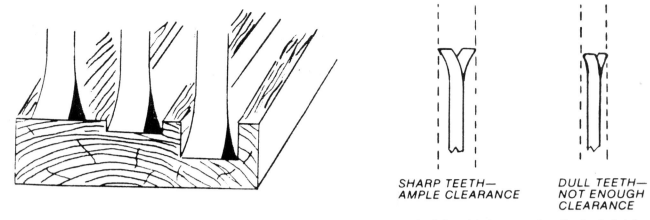

SHARP TEETH— AMPLE CLEARANCE

DULL TEETH— NOT ENOUGH CLEARANCE

Illus. 47 (above left). The slot or cut formed by the saw blade is called the kerf. Saw-blade set makes the kerf slightly wider than blade thickness. Illus. 48 (above right). The bend in the teeth is the set. Sharp teeth have more set than dull teeth. The set allows clearance for the blade to travel through the kerf.

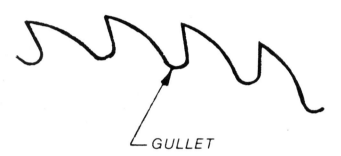

Illus. 49. The area behind the cutting edge of the tooth is the gullet. The gullet, which removes sawdust or chips from the kerf, is curved to reduce strain on the blade.

edge as it relates to the center line of the blade (Illus. 50). Rip blades usually have a hook angle of 30 degrees. Crosscut blades usually have less hook than rip blades. This is because excessive hook can cause tearout when crosscutting. Pull-stroke mitring machines work best when blades with a 0-degree hook are used because they have a tendency to "climb" the work when the blade has a hook angle of over 15 degrees. This climbing action could throw the entire machine out of adjustment (Illus. 51).

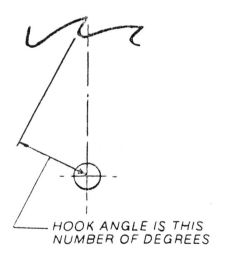

Illus. 50. The hook angle is measured between the centerline of the blade and the face of the tooth. Blades with large hook angles bite into the wood and remove large chips.

Illus. 51. These two saw blades have different hook angles. The woodworker II (front) is used on pull-stroke mitring machines. The woodworker II is used on chop-stroke mitring machines and portable circular saws.

Negative hook angles (Illus. 52) are sometimes used on blades intended for tough cutting jobs. Some circular-saw blades designed to cut used lumber have a negative hook angle. This allows them to cut nails or other metal in wood.

Be careful when cutting used lumber. Observe all

Illus. 52. Blades used for tough cutting jobs sometimes have a negative hook angle, which reduces dulling or tooth breakage when the blade hits a nail or other obstruction.

safety procedures and protect your eyes. Pieces of metal can be thrown at you by the blade.

Top Clearance (Illus. 53) is the downwards slope of the back of the tooth. This slope keeps the back of the tooth from rubbing on the wood. Without top clearance, the blade cannot cut.

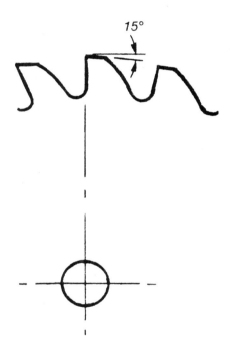

Illus. 53. Top clearance is the downwards slope of the back of the tooth. This clearance angle keeps the back of the tooth from rubbing or pounding on the wood as the cut is made.

The *arbor hole* is the hole in the blade that is used to mount it on the machine. Most arbor holes are circular, but some have a diamond shape. The diamond-shape arbor hole is commonly used on worm-drive portable circular saws. It ensures positive drive. There is no chance of slipping.

Some arbor holes have knockouts or bushings in the arbor hole which are designed to make the blade fit a number of different arbors. For example, a blade with a ⅝-inch arbor hole and a diamond knockout can be mounted on any arbor with a ⅝-inch diameter. If you remove the knockout the blade will fit a diamond arbor (Illus. 54). Most hardware stores sell knockouts or

Illus. 54. Diamond-shaped arbor knockouts are removed when the saw blade is to be used on a worm-drive portable circular saw.

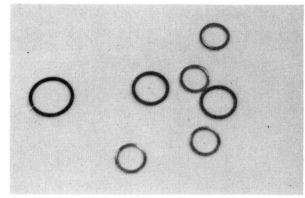

Illus. 55. Extra knockouts or arbor bushings are available at most hardware dealers. They are used to reduce the size of the arbor hole.

bushings that adapt the arbor hole to a different-size arbor (Illus. 55).

Work carefully when driving out a bushing or knockout. Excessive force could bend or damage the blade.

It is best to select a blade with the correct size and shape arbor hole to begin with. When bushings are added, they may not be concentric with the blade's arbor hole. This can cause the blade to orbit in an eccentric motion.

If a bushing must be used, it should be held securely in position. Use a prick punch and a ball-peen hammer to hold the bushing in place. Set the blade on a piece of scrap. Place the bushing in the arbor hole and offset the metal of the blade around the arbor hole with the prick punch and hammer (Illus. 56).

Illus. 56. A prick punch can be used to offset metal around the arbor hole. It holds the knockout in more securely.

When mounting any blade with a bushing, make sure that it is not knocked out by the arbor during the mounting process.

Expansion slots are found between the teeth on most blades. The high speed of the blade causes it to get hot when it is pushed into the work.

These slots allow the blade to expand as it heats up and keep it from warping. Some expansion slots have a soft metal plug at their bottoms (Illus. 57). The plug absorbs heat and reduces the noise level of the blade.

Illus. 57. The soft-metal plug in the expansion slot absorbs heat and reduces blade noise.

Removing or tampering with the plugs could adversely affect the blade's operation.

Newer blades have laser-cut expansion slots (Illus. 58). Blades with laser-cut expansion slots are less likely to crack under stress. Laser-cut expansion slots also reduce the chances of teeth bending.

Illus. 58. These expansion slots have been laser cut. Laser-cut slots are less likely to crack under stress.

Common Blade Types

Rip Blades Rip blades have deep gullets and large hook angles (Illus. 59–61). They have a straight-cutting edge designed to cut with the grain. This cutting edge looks like a chisel. Rip teeth are usually quite large.

Crosscut Blades Crosscut blades have smaller teeth than rip blades (Illus. 59–61). The teeth on a crosscut blade come to a point, not an edge. This allows them to cut the stringy fibres in wood with little pounding or tearing. A rip blade would pound and tear the fibres if it were used for crosscutting.

Most mitring machines will use some type of crosscut blade. This is because the teeth are fine and do not tear the wood even when cutting mitres.

Combination Blades Combination blades (Illus. 59–61) are designed for both ripping and crosscutting. They work very well for cutting wood fibres at an angle (mitre joints). Some combination blades have teeth that come to a point, but have a rip-tooth profile. Others have a chisel edge and a smaller hook angle. These blades do not produce smooth cuts, but they are well-suited for general carpentry or rough construction. They are used on portable circular sawing machines, and for most jobs the cut is adequate.

Smooth-cutting combination blades are sometimes called novelty combination blades. These blades have both rip and crosscut teeth. Novelty combination blades (Illus. 60 and 61) are preferred for cabinet and furniture work because they cut smoothly with little tear-out.

There are some hybrid combination blades being marketed today that are alleged to do all jobs. Those marketing the blade claim that they are the only blade needed. Most of these blades have been made to industrial standards, and are very high in quality. They will do 90% of the work, but there are some jobs they cannot do. There are too many compromises in the design of a saw blade to expect it to do all jobs well.

Hollow-Ground Blades Hollow-ground blades (Illus. 62 and 63) are blades with no set. The sides of the blade are recessed for clearance in the kerf. Some hollow-ground blades have sides that are recessed all the way to the hub (Illus. 64). Others are recessed only part of the way (Illus. 65). Blades with partially

Illus. 59. Shown here are three tool-steel blades. From left to right, they are a rip, a crosscut, and a combination blade. Note the Teflon coating on the combination blade. Its purpose is to minimize pitch buildup.

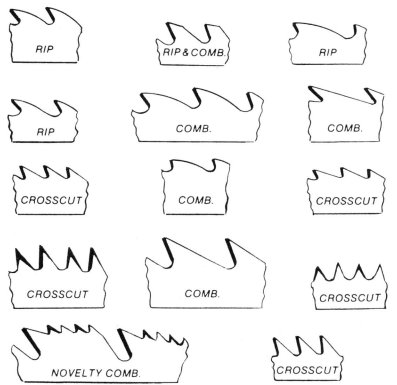

Illus. 60 and 61 (following page). Not all tooth shapes are as easy to identify as those in Illus. 59. Compare other blade profiles with the ones shown here and in Illus. 61 for identification.

BLADES:

A dull blade will cause slow, inefficient cutting and an overload on the saw motor. It is a good practice to keep extra blades on hand so that sharp blades are available while the dull ones are being sharpened. (See "SAWS—SHARPENING" in Yellow Pages). In fact, many lower-priced blades can be replaced with new ones at very little cost over the sharpening price.

Hardened gum on the blade will slow down the cutting. This gum can best be removed with trichlorethylene, kerosene or turpentine.

The following types of blades can be used with your saw:

COMBINATION BLADE—This is the latest-type fast-cutting blade for general service ripping and crosscutting. Each blade carries the correct number of teeth to cut chips rather than scrape sawdust.

CHISEL-TOOTH COMBINATION—Chisel-tooth blade edge is specially designed for general-purpose ripping and crosscutting. Fast, smooth cuts. Use of maximum speed in most cutting applications.

FRAMING/RIP COMBINATION—A 40-tooth blade for fascia, roofing, siding, sub-flooring, framing, form cutting. Rips, crosscuts, mitres, etc. Gives fast, smooth finishes when cutting with the grain of both soft and hard woods. Popular with users of worm-drive saws.

CROSSCUT BLADE—Designed specifically for fast, smooth crosscutting. Makes a smoother cut than the Combination Blade listed above.

RIP BLADE—Fast for rip cuts. Minimum binding and better chip clearance given by large teeth.

PLYWOOD BLADE—A hollow-ground, hard-chromed surface blade especially designed for exceptionally smooth cuts in plywood.

PLANER BLADE—This blade makes both rip and crosscuts. Ideal for interior woodwork. Hollow ground to produce the finest-possible saw-cut finish.

FLOORING BLADE—This is the correct blade to use on jobs when occasional nails may be encountered. Especially useful in cutting through flooring, sawing reclaimed lumber and opening boxes.

METAL-CUTTING BLADE—Has teeth shaped and set for cutting aluminum, copper, lead, and other soft metals.

FRICTION BLADE—Ideal for cutting corrugated, galvanized sheets and sheet metal up to 16 gauge. Cuts faster, with less dirt, than abrasive disc. Blade is taper-ground for clearance.

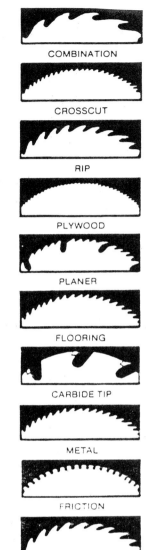

Illus. 61.

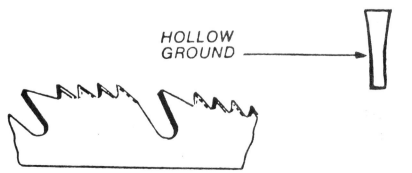

Illus. 62 (far left). A hollow-ground blade has sides that are relieved or ground thinner than the teeth. Illus. 63 (left). The thinner sides of the blade allow the teeth to cut without set. A hollow-ground blade does not cause as much tear-out as a blade with set.

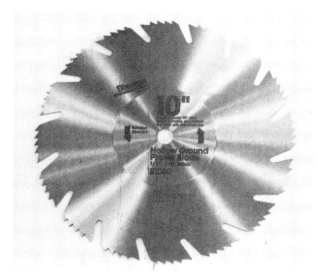 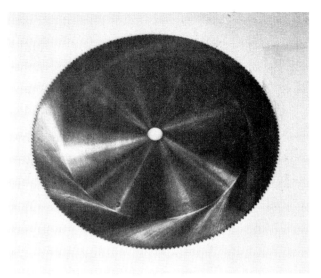

Illus. 64 (above left). This hollow-ground blade has its sides ground back to the hub. It will cut through thick stock without rubbing on the kerf sides. Illus. 65 (above right). The sides of this hollow-ground blade are only ground part of the way back to the hub. It is designed to cut through sheet and solid stock less than 1¼ inches thick. This blade has greater rigidity than those ground all the way to the hub.

recessed sides cannot cut thick stock, but are more rigid.

Hollow-ground blades cause less splintering and tear-out in the wood they cut. The sides of the blade may burn and accumulate pitch (wood residue) if they are used for heavy cutting instead of finish-cutting. Hollow-ground blades work best with very true stock. Use them to cut mitres and compound mitres. Avoid using them for heavy ripping. Most hollow-ground blades have novelty combination teeth.

Hollow-ground blades are sometimes called *planer* blades (Illus. 61) because the wood is so smooth after being cut that it appears to have been planed. There are also blades with abrasives attached to each side. These blades sand the sides of the kerf after the blade cuts the wood.

Plywood Blades Plywood blades, sometimes called panelling or veneer blades, are designed to cut hardwood plywood with cabinet-grade or furniture-grade outer veneers. These blades have very fine crosscut teeth with little set (Illus. 66). Some of these blades are hollow-ground. The fine teeth and small amount of set allow very smooth splinter-free cuts.

These blades should be used only when appropriate. Using them for other purposes can ruin them quickly. Certain types of plywood cores (particle or fibre) can dull these blades quickly. Carbide-tipped blades would be a better choice for particle- or fibre-core plywood or other sheet stock.

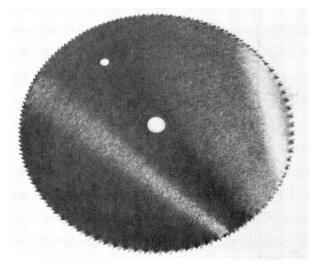

Illus. 66. This plywood blade has fine teeth and very little set. Use these blades for finish cuts only. In general use, the blade will bind in the kerf and dull quickly.

Triangular Blade A specialty blade known as the Thorsness™ blade has a triangular shape (Illus. 67). This blade is designed for specialty cutting. It will rip or crosscut the same way as any other blade. In addition, it disengages with the work three times per revolution, which allows it to make irregular or curved cuts (Illus. 68).

The triangular blade is made from tool steel. It is

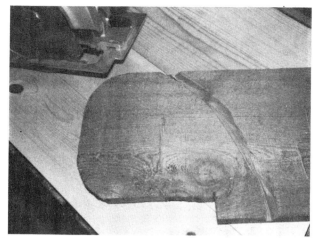

Illus. 67 (above left). This Thorsness blade is a specialty blade. In addition to conventional cuts, it will also cut curves and dadoes. Illus. 68 (above right). The triangular shape of this blade provides enough clearance so that it can make a curved cut.

not hard enough for work in particle board or other composition materials. The hardness of the glue in the particle board dulls the steel rapidly. The triangular blade is designed for use on a portable circular cutoff saw or a table or radial-arm saw.

Chain-Saw Blade The chain-saw blade is a specialty blade that is a combination of a circular-saw blade and a chain-saw blade. The circular-saw body has a saw chain wrapped around its circumference (Illus. 69). As the saw turns, the chain does the cutting. This blade is marketed as the Beaver™ Blade.

Illus. 69. This specialty blade has a chain-saw blade wrapped around it. This blade will make any conventional cut.

The chain-saw blade can be used on portable circular cutoff saws and table and radial-arm saws. Its kerf is about twice that of a conventional circular-saw blade (Illus. 70); this reduces the chance of kickback because the blade does not bind as easily. The larger kerf also makes the blade useful for cutting wood with a high-moisture content. The chips clear easily and there is little chance of binding.

The chain-saw blade is coarse, but it leaves a relatively smooth cut. The larger kerf requires more energy for cutting, so the feed is not as fast as one might expect. Again, the cutting tips on the chain are not hardened, so they should not be used to cut particle board. This would cause rapid dulling.

The chain-saw blade is sharpened and maintained the same way as a saw chain. Inspect the blade periodically to be sure that the links are tight and the teeth are sharp. Use a professional sharpening service for the best results.

Metal- and Masonry-Cutting Discs Portable circular cutoff saws are sometimes used to cut metal or masonry. The discs used for these jobs do not have teeth like a conventional saw blade. They look more like a grinding wheel, and cut in the same fashion. These discs are usually ⅛ inch thick. They are designed for use only on a portable circular cutoff saw (Illus. 71).

The discs used for cutting masonry are usually made from silicon carbide. Those used for metal are usually made from aluminum oxide. When using any metal- or masonry-cutting disc, be sure to follow the manufacturer's instructions. Some metal-cutting discs

Illus. 70. The chain-saw blade makes a kerf about twice as wide as a conventional blade; there is less chance of kickback with a wide kerf. The chain-saw blade cuts green (wet) lumber efficiently.

Illus. 71. The silicon carbide disc shown on the left is designed for cutting masonry. Always check the manufacturer's instructions for operation and safety practices. Note: Some of these blades have to be used with a coolant. The aluminum oxide disc on the right is designed for cutting metal. Make sure that the disc is designed for use on the metal you intend to cut, and that the disc is rated for the speed of the saw.

are not intended for certain metals, and some masonry discs cannot be used without a water coolant.

Make sure that the saw does not exceed the safe speed of the cutting disc. Some discs are not compatible with a high-speed saw (revolutions per minute). Also, make sure that the drive is correct. Some discs cannot be used with a diamond-shaped arbor hole drive.

Diamond Blades Diamond blades are metal discs with diamond particles bonded to their sides and peripheries. The blades may be solid or segmented. Some are intended for specific materials such as glass; some of these blades require coolant during operation.

Be sure to read the manufacturer's instructions and recommendations for use. If you use a coolant, control the amount of water near the work area. Large puddles of water can increase the chance of electric shock.

Carbide-Tipped Blades Carbide-tipped blades have teeth made from small pieces of carbide. The carbide is brazed onto the circular blade. Usually, there is a little seat cut in the blade; this is where the carbide is brazed. Most carbide tips are wider than the metal blade, so no set is required. Carbide-tipped blades stay sharp 5 to 10 times longer than conventional blades.

Carbide is much harder than the steel used for conventional blades. Because of its hardness, carbide is also quite brittle; it will fracture easily if struck against a hard object. Carbide-tipped blades must be handled with care.

Carbide-tipped blades are more expensive than steel blades, but they require much less maintenance. They are preferred for tough materials such as hardboard, plastic laminate, and particle board.

Carbide-tipped blades come in rip, crosscut, combination, hollow-ground, and plywood categories. They do not always resemble their steel counterparts. Usually the type of teeth, the number of teeth, the hook angle, and top bevel angle determine the blade's function (Illus. 72–74). They can have alternate top bevel, triple chip, rip, cutoff, or combination teeth (Illus. 75).

Illus. 72. This carbide-tipped blade is a rip blade. It has a high hook angle for removing stock quickly.

Illus. 73. This 40-tooth carbide-tipped blade is designed for general cutting on a 10-inch chop-stroke mitring machine.

Illus. 74. This 48-tooth carbide-tipped blade is designed for cutting sensitive materials such as hardwood plywood on a pull-stroke mitring machine.

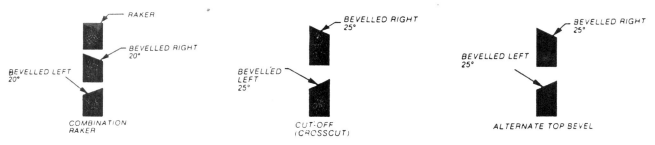

75. Some common carbide-tooth shapes and groupings. Rip blades have flat raker teeth only. Combination blades have rakers ground slightly lower than the other teeth. This allows them to clean out the kerf without causing tear-out. Alternate top bevel blades are used for fine cutting of veneers where tear-out could be a problem.

Some carbide blades are called *control-cut* blades. These blades have 8–12 teeth, which are set slightly above the blade's periphery. These teeth minimize the chance of a kickback or severe cut. Consult a manufacturer's catalogue for specifics and blade function (Illus. 76).

Illus. 76. The teeth on this Credo® saw blade extend slightly beyond the saw rim. This limits and controls the cut.

More contemporary carbide tooth design has changed the ground shape of the carbide teeth. These shapes improve cutting under certain conditions. The Piranha® blade, which is marketed by Black and Decker, has a curved face on the carbide tip (Illus. 77) that promotes faster and smoother cutting. In addition, this blade is ground so that the non-cutting side of the carbide tip does not drag on the saw kerf (Illus. 78). Teeth ground to this configuration are sharpened on the top only (Illus. 79).

Another blade manufactured by Vermont American has a curve on the face of the tooth. It is referred to as the Particleboard/Plywood Blade. This blade is sharpened in the same way as the Piranha® blade.

The Laser X2™ is another blade with a new tooth configuration. The face of the tooth is V-shaped. This means that each tooth has two top bevels. The V

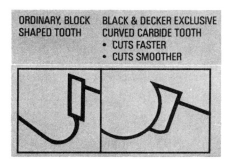

Illus. 77. The curved face on this carbide tip promotes faster and smoother cutting.

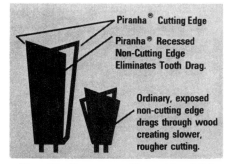

Illus. 78. The non-cutting side of this carbide tip does not drag on the saw kerf. This makes cutting faster with less heat buildup.

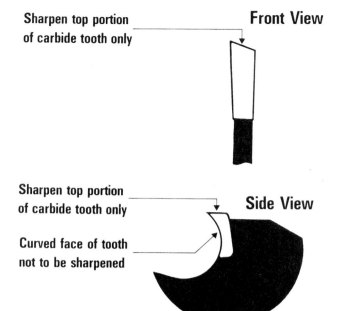

Illus. 79. The Piranha™ tooth is sharpened only on the top.

remanufactured wood products such as plywood, particle board, and hardboard.

Evaluating Carbide Blades Not all carbide blades do the job equally well. Before buying any, look them over and evaluate them carefully. The size of the carbide tips is important (Illus. 84). The larger the tips, the more times the blade can be sharpened. Look at the braze joint between the blade and the carbide tip. It often indicates blade quality. There should be no pits in the braze.

Inspect the teeth. They should be ground smooth

Illus. 80. This Cuda blade has the carbide tips mounted horizontally, which eliminates tooth drag and blade noise.

shape promotes balanced cutting, reduces splintering, reduces blade drag, and enhances the life of the blade. The Laser X2™ is manufactured by Vermont American, and can also be resharpened.

Credo®'s 'Cuda™ Blade is an ultra-thin carbide-tipped blade. The carbide tips are mounted horizontally instead of vertically (Illus. 80). The kerf that this blade produces is noticeably thinner than those made by conventional blades (Illus. 81). You need less energy to cut the kerf and can feed the blade into it more quickly. The thinner blade and kerf also reduces blade noise.

Carbide-tipped blades have different designs for different saws or jobs. Generally, those blades used in mitring machines have a higher alternate top bevel and a smaller kerf than those used on a portable circular saw. The higher alternate top bevel makes the bottom of the mitre cut smoother. A larger kerf on portable circular cutoff machines helps reduce blade binding, which is a common cause of kickback.

Laser Blade® The Laser Blade is a carbide-tipped blade with an unconventional shape (Illus. 82). It can be used on any portable saw for common cuts, and for specialty cuts such as curves, dadoes, rabbets (Illus. 83).

The carbide tips on this blade allow it to be used on

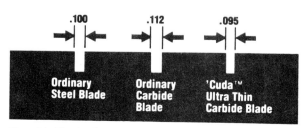

Illus. 81. The kerf produced by the Cuda blade is noticeably thinner than those produced by conventional saw blades.

Illus. 82. This carbide-tipped blade has an unconventional shape. Note that all teeth are not following the blade's periphery. This blade is known as the Laser blade. It will make common and specialty cuts.

Illus. 83. The Laser blade will cut rabbets, dadoes, and curves. Since it has carbide tips, it can be used on particle board and other tough materials.

Illus. 84. Can you see any difference between these two carbide-tipped blades? Do you think these blades should sell for the same price?

(Illus. 85). The smoother the surface of the carbide, the better the cut and the longer it will stay sharp.

Keep the carbide blade sharp. Use a reliable sharpening service that polishes the teeth and leaves no coarse grinding marks. Blades that are ground using a coolant will usually be smoother and sharper than those which are ground dry.

Selecting Blades

One blade cannot perform all sawing operations. There are differences in the blades used on mitring machines and portable circular-sawing-machines. For the 7¼-inch portable circular saw, three blades will do most jobs. They are as follows:

Rip blade: 8–16 teeth
General-duty blade: 24 teeth
Fine-work blade: 40 teeth.

When selecting a blade for use with your portable circular saw, use the following general rules to help you select the correct one:

1. Three teeth should be in the work at all times.
2. Larger teeth are best for ripping.
3. Use a rip blade when the job is strictly ripping.
4. Small teeth mean a smoother cut and a slower feed rate.
5. Use the largest-tooth blade that will produce satisfactory results.
6. Smaller-diameter blades have more power because less energy is required to turn them.
7. The thinner the kerf, the less energy needed; there is, however, an increased chance of blade flutter.
8. Hollow-ground and panelling blades should be used only for true, dry cabinet-grade lumber.
9. Remove high-quality or specialty blades as soon as the job is done.
10. Green lumber and construction lumber require blades with more set than dry hardwood lumber. This is due to the increased moisture content.
11. Never use a dull blade. It is unsafe and produces poor results.
12. Do not use a tool-steel blade on particle board, modified wood products, or treated lumber.

Analyze the job and mount the correct blade on your saw. Remember, the time spent changing blades is time well spent. The correct blade does the most efficient and safest job. Dull blades waste time and energy.

Through trial-and-error, you will learn to select the best blade for every job you do. Make note of which blade does the best job. This provides a ready reference for future use.

Most chop-stroke mitring machines use a finer blade than a portable circular cutoff saw. Generally, mitring machines with a blade of 7–10 inches in diameter use a blade with 40–80 teeth. Some 10-inch machines use the same blade that is used on a table or radial-arm saw for fine cutting. Make sure that these blades can handle the higher speed (rpm's) of the mitring machine. Most table and radial-arm saws rotate at a lower speed than a mitring machine.

When working with chop-stroke mitring machines,

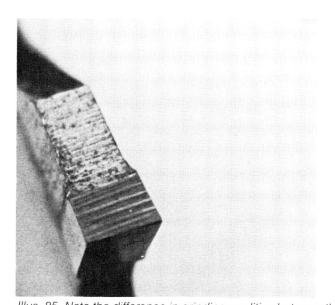 

Illus. 85. Note the difference in grinding qualities between the rough surface (left) and smooth surface (right). Smoother surfaces mean sharper teeth and greater longevity.

remember that the angle at the alternate top bevel on each tooth is important. The higher the bevel angle, the better it will cut a mitre across the grain fibres. This is particularly true when working with finished trim or specialty materials. The higher alternate top bevel makes the carbide on the tooth weaker. Inspect these blades periodically for dulling or tip breakage.

Blades used on compound-cut saws may be fine or coarse, depending on the operation. Generally, compound-cut saws use blades 7–8 inches in diameter. Blades with 40–60 teeth are used for fine work, and blades with 24 teeth are used for coarse work. When using a compound-cut saw, remember that the hook angle is as critical as the number of teeth on the blade. Blades with a hook angle of 15 degrees or more tend to climb the wood; this can damage the saw or blade and make the cut appear rougher than it should be.

Blade Maintenance

Blades should be protected from damage when not in use. The teeth of blades in storage should not touch. Such contact can dull or break carbide teeth, and will dull steel blades. Hang blades individually or with cardboard spacers between them (Illus. 86). This will keep them sharp. Protect blades from corrosion, which will deteriorate a sharp cutting edge.

Handle circular-saw blades carefully. A sharp (or dull) blade can cut you. Never lay a blade on a metal surface. The set of the teeth wears against the metal and becomes dull.

Pitch When a circular saw blade becomes hot, pitch will accumulate on it. Pitch is a brown, sticky substance (wood resin) that looks like varnish or gum. As pitch accumulates on the blade, it acts as an insulator; it keeps the blade from dissipating heat and causes it to become dull faster.

Pitch is usually a sign of a blade with too little set for the job. It can also mean that the blade is too dull to cut. In some cases, the blade accumulates pitch and smokes when it is installed backwards (teeth pointing the wrong way). Some blades are Teflon™ coated to resist pitch accumulation, but the Teflon™ on some blades may wear off after two or three resharpenings.

Commercial pitch removers can be used to clean blades. Kerosene, liquid hand cleaners (Illus. 87), hot water, and oven cleaners also work well. Avoid using abrasives to remove pitch. Abrasives leave scratches that make it easier for pitch to anchor itself to the blade.

Pitch accumulation does not always mean the blade is dull. A sharp hollow-ground blade will accumulate pitch quickly when used for heavy ripping.

Dull Blades Some indications of a dull blade include the following: 1, the blade tends to bind in the cut; 2, the blade smokes or gives off a burning odor; 3, increased effort is needed to feed the stock into the blade; and 4, the saw no longer cuts in a straight line.

You can also identify dull blades by looking at them. Look at the teeth. Rip teeth should come to an edge (Illus. 88). The edge should be a straight line and not rounded. Crosscut teeth should come to a point. The two cutting angles should form a straight line to the point of the tooth (Illus. 89).

Carbide teeth stay sharp longer than steel teeth, but they also become dull. If a dull carbide blade is left on the saw, the brittle teeth will crack or shatter. To determine if carbide teeth are sharp, drag your fingernail across the carbide tip (Illus. 90). It should cut a chip (remove a curl from your fingernail). If it does not (fingernail slides across the tip), it is too dull to cut properly.

Illus. 86. Protect the cutting edges of your saw blades with cardboard spacers. This should also minimize the chances of chipping the carbide tips.

Illus. 87. Liquid hand cleaners will remove pitch from saw blades. They are safer than most chemicals, and are water soluble.

Illus. 88 (above left). The rip teeth shown here have rounded ends. This means that they are dull. The ends should form a straight line or edge. More energy and feed pressure is needed to make a rip cut with a dull blade. Illus. 89 (above right). Crosscut teeth also become rounded or flat on their ends. A sharp crosscut tooth comes to a point.

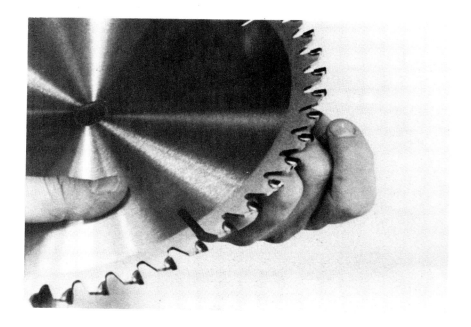

Illus. 90. If a carbide-tipped blade is sharp, it will raise a chip on your finger if you drag your fingernail across the carbide tip (you can see one on my index finger). If it is dull, your fingernail will slide across the tip.

Disconnect the saw to check a blade that is mounted. Replacing broken carbide tips is much more expensive than sharpening. Keep carbide blades sharp, and broken tips will not be a problem.

Getting Blades Sharpened It is best to have your blades sharpened by a professional. The equipment they use is very accurate, but very expensive (Illus. 91). Find a reliable service and develop a good working relationship.

Not all sharpening services are equal. Some do better work than others (Illus. 92). When trying a new service, do not send them your best blades. Have them sharpen one or two general-duty blades first. Inspect the blades carefully (Illus. 93). If the results are not satisfactory, try another sharpening service.

Make a board for transporting blades (Illus. 94). Put cardboard spacers between the blades. This will make the blades safer and easier to transport. It will also keep them well-protected and sharp.

Illus. 91. Professional sharpening equipment is very accurate, but too expensive for the average woodworker. A quality sharpening job is a bargain. Note the water coolant hose.

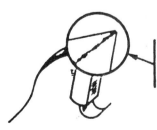

Edge produced by most saw sharpening services (not visible to the naked eye)

Illus. 92. Water coolant and fine grinding wheels produce a very sharp carbide tip. The water also reduces heat stress on the braze joint.

Edge produced by our superior factory methods and equipment

Illus. 93. Inspect your blades after they have been sharpened. When the tips have been ground as smoothly as the ones shown here, the blade will remain sharp for a long time.

Illus. 94. A sharpening or transportation board like this will help keep blades sharp. Its presence will also suggest to the person who sharpens your blades that a good job of sharpening is expected.

Hand-Held Circular Saws

4

Introduction to the Portable Circular Saw

Saw Size

The sizes of portable circular saws are determined by blade diameter and horsepower. The blades for portable circular saws range in diameter from approximately 4–16 inches (Illus. 95). The most common size seems to be 7¼ inches (Illus. 96). When a 7¼ inch blade is tilted at a 45-degree angle, it will cut through stock 1½ inches thick. This means that the common portable circular saw can cut a two by four (1½ × 3½ inches) in half when the blade is tilted at a 45-degree angle (Illus. 97).

Smaller-diameter saws are usually used for trim work (Illus. 98). These saws, which have less power than a 7¼-inch saw, are used for cutting jamb stock, moulding, flooring, and roof sheathing (Illus. 99). Some do not accept accessory blades for cutting metal or concrete. Consult the owner's manual to be sure the saw is designed for the job you intend to do.

Portable circular-saw horsepower can either be actual or developed. Actual or rated horsepower is the saw's power under load (blade is cutting). Developed horsepower is the saw's power with no load (blade is turning, but not cutting). Most saws range in horsepower from ½ to 2½.

Since the terms actual and developed horsepower are often confused, it is best to compare saws by ampere ratings. Generally, ten amperes is roughly equal to one horsepower. Two horsepower is roughly equal to 15 amperes. Portable circular saws rarely draw over 15 amperes because most common electrical circuits are rated at 15 amperes. Overloading a circuit can trip a circuit breaker or fuse.

Operating a portable circular saw below the correct amperage can damage the motor. Extension cords that are too small can also cause amperage problems. Consult the electrical information section in Chapter 2 to make sure that the extension cord will handle the amperage. Generally, a 7¼-inch saw should have a one-horsepower (10 ampere) motor or larger.

Three other common determiners of saw size are weight, base size, and arbor speed. Weight varies among portable circular saws. Obviously, a 16-inch saw is much heavier than a 4-inch saw, but saws of the same blade size also vary in weight.

Though the weight difference between two saws may be slight, it becomes more evident after eight hours of constant use. A 7¼-inch worm-drive saw weighs more than a 7¼-inch direct-drive saw (Illus. 100), and the weight of direct- and worm-drive saws varies. Some woodworkers consider weight first when buying. Others use weight only as a minor consideration.

Base size is important when considering the size of the saw (Illus. 101). Some bases are much larger than others on the same-sized saw. A large base makes the

Illus. 95. Portable circular-saw diameters range from 4 to 16 inches. The job dictates the saw you select.

Illus. 96 (above left). The most common portable circular saw uses a 7¼-inch blade. This size seems most popular with woodworkers. Illus. 97 (above right). A 7¼-inch saw can cut a 2 × 4 in half when tilted at a 45° angle. Today, some smaller saws can do the same job.

 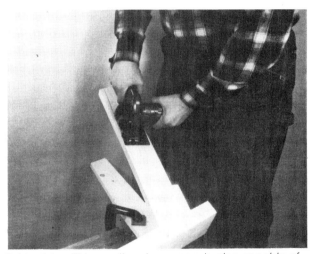

Illus. 98 (above left). Small portable circular saws are used on lighter jobs. This smaller plunge saw is also capable of cutting framing stock. Illus. 99 (above right). Smaller saws are used for trim work and for cutting flooring, sheathing, and jamb stock.

Illus. 100. The 7¼-inch worm-drive saw (right) weighs more than the direct-drive saw. Note also that the blades of these two saws are on opposite sides.

Illus. 101. Note the differences in saw base sizes. Large bases mean better saw control but poor maneuverability.

saw easier to control, but makes it awkward in tight spots.

Arbor speed affects the blade speed in revolutions per minute (rpm's). Common portable-circular saws (7¼ inch) have arbor speeds that range from 4,300–5,800 rpm's. The higher the arbor rpm's, the faster the feed. This is because the tip speed increases, and the teeth remove stock faster. Worm-drive saws generally turn slower than direct-drive saws.

Types of Power

The four most common types of power for a portable circular saw are electric, battery, pneumatic, and internal-combustion power. Most portable circular saws use 110-volt alternating current. This is the most common electrical circuit in the United States.

Some electric saws now have electronic components that control the motor (Illus. 102). These saws regulate the cutting speed. If the motor begins to slow down during a heavy cut, the electronic controls instantly boost the motor speed and power to compensate for the increased resistance (Illus. 103). These saws make it easier to cut rough stock because they are less likely to bind when they hit a knot or the kerf closes on them.

Illus. 103. The electronic controls give the operator feedback concerning correct cutting practices. This increases saw longevity and reduces sawing noise.

Electronic saws also have a soft-start feature that brings the blade to full speed slowly. This reduces saw torque in your hand and increases the life of the saw.

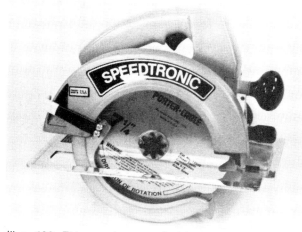

Illus. 102. This saw is controlled by electronic components. The electronic controls monitor load and speed. They keep the saw blade at optimum speed and power regardless of what is being cut.

An electronic saw will also cut straighter because blade speed remains constant. If the blade changes speed during a cut, it becomes more difficult to cut straight even when you are using a cutting guide or straightedge. Electronic controls also reduce sawing noise because speed control is more accurate.

Battery-powered portable circular saws have become very popular because of improved battery technology. Smaller, lighter batteries now possess more power and endurance. They also recharge in a much shorter period of time. The use of smaller, thinner blades has also made battery-operated saws more popular because these blades require less energy to cut with.

Battery-operated portable circular saws use blades as large as 6¼ inches in diameter (Illus. 104). The batteries are made of either lead–acid or nickel–cadmium (Ni–cad).

Illus. 104. Cordless saws use blades as large as 6¼ inches in diameter. Cordless saws are very powerful and should be treated like any other power saw.

When I compared three common brand-name 6¼-inch cordless saws, I discovered that one used a Ni–cad battery, while the other two used lead–acid batteries. The Ni–cad battery moved the blade at a lower speed (1,000 rpm's) than the lead–acid batteries (3,200 and 3,400 rpm's), but had a longer cutting time. All the batteries charged in about one hour.

Many carpenters buy two batteries so that one is always charged and ready for use. Most batteries are rated for 1,000 recharges. Cold temperatures will affect the battery's charge. For optimum efficiency, these batteries should be kept at room temperature and charged according to the manufacturer's instructions.

When charging batteries, observe correct polarity and avoid overcharging (Illus. 105 and 106). Make sure that the charger is correct for the battery. Using the wrong charger can damage the charger or battery. Charge the battery only after it has been completely discharged. Recharging a battery before it has been completely discharged will reduce its useful life and charging capacity.

Be sure to handle any cordless saw as if it could start any second. Remove the battery for adjustment or blade changing just as you would unplug an electric saw. A cordless saw is as dangerous as an electric saw—the electricity comes from the battery.

Use cordless saws when in the proper frame of mind. A cordless saw is not a toy. It is a powerful tool capable of a kickback or of cutting anything in its path.

Pneumatic portable circular saws are driven by compressed air. These portable circular saws resemble all other saws, except they have an air connection instead of a power cord (Illus. 107). The pneumatic portable circular saw has an air motor, which is an impeller-type motor. The compressed air turns the impeller, which drives the saw blade.

Pneumatic saws are favored by large industrial users. Although they cost more than saws with other

Illus. 105. Be sure to use the correct charger and observe correct polarity when charging batteries.

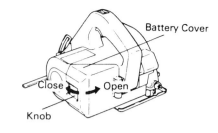
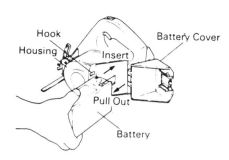

Illus. 106. Install and remove batteries properly when using cordless saws. Most batteries can only be inserted the correct way, so do not force them.

Illus. 107. This circular saw is driven by compressed air. The compressed air turns an impeller, which drives the blade.

power sources (about six times the cost of an electric saw), they have a few distinct advantages. First, they do not wear out like an electrical circular saw. The air turns the impeller, but causes little wear. The electrical motor wears first at the brushes and switch. This wear usually leads to saw failure.

Second, the pneumatic portable circular saw runs cooler than an electric motor, which gives off heat during operation. The pneumatic motor is cooled by the expansion of the compressed air.

Third, there is little chance of electrical shock while using the compressed-air portable circular saw. While electrical saws should not be used in damp or wet conditions, these conditions pose no hazard while using a pneumatic saw.

Fourth, the rpm's remain constant under load. The saw does not bog down during a heavy cut. This is desirable when doing work that demands accuracy.

Pneumatic portable circular saws are favored for abrasive cutting in metal and concrete products. This is because of the constant speed and its tolerance for metallic and stone particles.

Pneumatic saws require air pressure of 90 psi (pounds per square inch) and consume 80 cubic feet of air per minute. A large industrial-type compressor is needed to power a pneumatic portable circular saw. Many of these saws also require that the air be conditioned with a lubricating oil. This oil, which is mixed with the compressed air in droplet form, travels throughout the tool with the air and lubricates the internal parts. The oil is quite thin and is designed for use in pneumatic tools. Substituting other types of oil is not recommended. They can cause the seals inside the saw to deteriorate prematurely.

Internal-combustion saws use an engine similar to that used on a chain saw. These saws are not new, but the older models were less reliable. The newer models have generated interest due to their reliability and the fact that they can be used in the field when electricity is not available or where stringing an electrical cord is not practical.

Classifications

Portable circular saws can be classified in a number of different ways. Understanding these classifications will help you select a portable circular saw according to the job or your needs.

Worm-Drive or Direct-Drive Motor The motor on a portable circular saw can be parallel or perpendicular to the blade. When the motor is parallel to the blade, the drive type is known as worm-drive (Illus. 108–110). When the motor is perpendicular to the blade, the drive type is known as helical-gear drive, or direct drive (Illus. 111 and 112). Worm-drive saws have a gear mechanism (worm gear) between the motor and blade (Illus. 113). Direct-drive saws have an arbor that extends directly from the end of the motor or through a reduction gear. The blade is attached directly to the arbor.

Special helical gearing can make the direct-drive

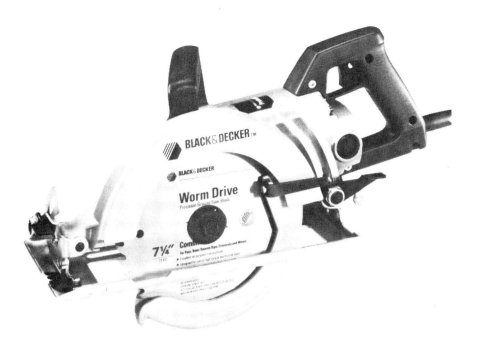

Illus. 108. The motor on a worm-drive saw is parallel to the blade. Worm-drive saws have a positive drive, so the blade does not slip or spin freely.

Illus. 109. These two worm-drive saws represent two of the most common sizes: 4½ and 7¼ inches.

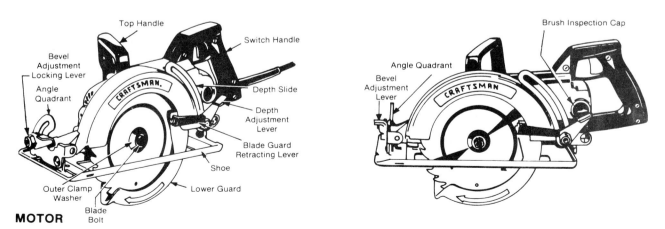

Illus. 110. Study these drawings to familiarize yourself with the parts on worm-drive saws.

Illus. 111. Direct-drive saws usually have an arbor that extends from the motor. The blade is usually perpendicular to the motor.

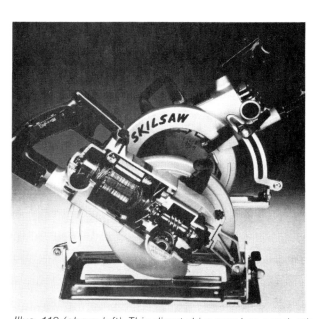

Illus. 112 (above left). This direct-drive saw has a reduction gear system that's similar to a worm-gear system. Study this photograph to see how the parts of a portable circular saw interact. Illus. 113 (above right). Can you find the worm gear in this photograph? The worm-gear area is filled with oil for operation.

saw very powerful. Direct-drive saws are usually lighter and more compact than worm-drive saws. Lightness is favored by some carpenters and woodworkers, but others maintain that the extra weight of the worm-drive saw makes it less likely to kick back and easier to use when cutting downwards. When you are cutting downwards, the weight of the saw does the cutting. No extra force is needed to make the cut.

Direct-drive saws are sometimes called "side winders" because the blade is attached to the side of the motor. Worm-drive saws are sometimes called "squirm drives" because of their tendency to turn slightly or "squirm" when started.

Handles The main handle on a portable circular saw can be a thrust handle or a top handle (Illus. 114). A thrust handle is located so that you can push the saw

Illus. 114A and B. These two saws are identical except for the handles. The thrust-handle (left) is situated for pushing through the cut. The top-handle (right) is positioned for more control and less thrust.

(thrust) as it cuts. The top handle is located on top of the saw.

For accurate work, the top handle offers greater control. When you are making heavy cuts, the thrust handle helps you push the saw through the work. It should be noted that some thrust handles become top handles when the depth of cut is reduced, but this happens only on pivot-base portable circular saws. There are other power saws with round handles that are gripped in a way that provides both control and thrust.

Right- or Left-Hand Saws The hand of the saw is determined by the side of the saw on which the blade is mounted. For example, when the saw is held normally and the blade is on the right side of the saw, it is a right-hand saw. Most direct-drive saws are right-hand saws, and most worm-drive saws are left-hand saws. However, there are exceptions (Illus. 115 and 116).

There are differing opinions about which saws (right-hand or left-hand) are easiest to use and control. Some right-handed workers prefer a right-hand saw, while others prefer a left-hand saw. The left-hand saw allows a right-handed operator to see the blade and cut without looking over the saw (Illus. 117). The right-hand saw is a little more awkward to use because the operator must look over the saw to see the blade and the cut (Illus. 118).

If you have enough skill and experience, you can use both the right-hand and left-hand saw. In fact, many experienced woodworkers can use a right- or left-hand portable circular saw equally well. If you are not experienced with either saw, a few practice cuts in scrap stock can help you get the feel of the saw.

Illus. 115. Note the left-hand (right) and the right-hand (left) direct-drive saws. Which would be easier for you to use?

Illus. 116. This worm-drive saw is a right-hand saw. It is the only right-hand worm-drive saw in production.

Illus. 117 (above left). The left-hand saw allows the right-hand operator to see the blade and cut without looking over the saw. Illus. 118 (above right). The right-hand operator has to look over the right-hand saw while cutting. Some people feel that this is awkward.

Bases The base or shoe on the portable circular saw rides on the work while the cut is being made. The blade protrudes through the base to make the cut. To control the amount of blade exposure, move the base up and down. The way the base is moved determines its classification.

The base on a portable circular saw can be a tilt base or parallel base (Illus. 119). The tilt base is also known as a pivot shoe. The parallel base is also known as a drop shoe. A tilt-base saw has a point at the front or back of the saw on which the base pivots. A wing nut locks the adjustment at the opposite end of the saw (Illus. 120). All worm-drive saws and many direct-drive saws use the pivoting base.

Some direct-drive saws have a parallel base. The parallel base slides on machined ways near the front of

Illus. 120. The pivot base pivots from the front of the base. The wing nut at the rear of the saw locks the adjustment.

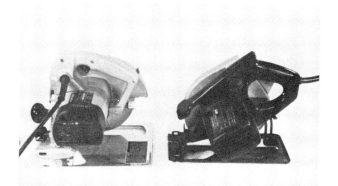

Illus. 119. The parallel base (left) and the pivot base (right) control blade exposure differently.

the saw (Illus. 121). A threaded lock knob holds the setting after adjustment. This knob also serves as a secondary control handle when the saw is cutting.

When the saw has a tilting base, the balance of the saw changes when you change blade exposure. The less blade exposed, the higher the weight of the motor if lifted. This can make the saw more difficult to control as you make the cut. The parallel-base saw keeps the motor parallel to the base as you reduce blade exposure. This makes the saw feel more balanced when you use it. With experience, the operator can control either type of saw well, but the novice will find the parallel-base saw easier to control.

Illus. 121. The parallel-base control consists of machined ways that are attached to the base. The base moves up and down on the ways. The large knob locks the adjustment and provides additional control.

Illus. 123. Direct-drive saws usually employ a round arbor hole and a spring-loaded arbor washer. The spring-loaded arbor washer allows the blade to stall while the motor keeps running. This minimizes the chance of kickback.

Positive Drive or Clutch Drive A portable circular saw can be either positive drive or clutch drive. A positive-drive saw has a diamond-shaped arbor. The blade has a mating diamond-shaped arbor hole. This engagement eliminates slipping of the blade on the arbor. The blade drive is positive. Worm-drive saws employ a positive-drive system (Illus. 122).

Illus. 122. The worm-drive saw employs a diamond drive on the saw blade. The diamond drive is a positive engagement. The blade does not stall in the wood unless the motor stalls.

Most direct-drive saws use the clutch-drive system. The arbor is round, as is the hole in the blade. A spring washer is used between the outer arbor washer and the arbor nut (Illus. 123). This allows the arbor to slip if the blade becomes jammed or pinched in the work.

Clutch-drive saws are less likely to kick back than a positive-drive saw. When the blade slips, a kickback does not usually occur. Positive-drive saws are designed for absolute power and tough-cutting jobs. The blade does not slip and can kick back if the blade becomes pinched or jammed. Keep your hands clear of the blade's path (front and back). This will minimize the chance of contact with the blade in the event of a kickback. Be sure to read and observe the safety practices in this chapter before operating any portable circular saw.

If the blade on a clutch-drive saw binds and slips, retract the saw slightly while you release the trigger switch. Eliminate the cause of the binding, turn on the saw, and resume cutting.

Plunge-Cutting or Regular-Cutting The regular-cutting portable circular saw has a lower telescoping guard. The guard is lifted by the work as you make the cut. This is the most common type of portable circular saw.

Some plunge-cutting portable circular saws have no lower telescoping guard (Illus. 124), while others do. When you use the saw, the blade plunges from the base of the saw to make the cut. The plunge-cutting saw has a release mechanism that allows the blade to drop through the base (Illus. 125). Blade exposure is controlled by a mechanism on the saw.

The plunge-cutting portable circular saw with no lower guard reduces drag on the work. This drag can affect the accuracy of the cut, and may cause the saw to turn in the cut. As soon as you release pressure on the control handle, the blade retracts into the housing or the lower guard returns. If a kickback occurs, the blade actually tries to lift itself into the housing or is protected by the lower guard.

The plunge-cutting portable circular saw is a rela-

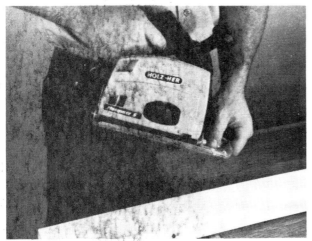 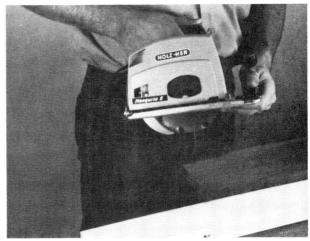

Illus. 124 (above left). There is no lower telescoping guard on this plunge-type saw. The blade plunges through the base when being used for cutting. Illus. 125 (above right). The thumb releases the plunge mechanism and the blade drops through the base. A stop controls blade exposure. Note the splitter attached to the motor housing. It keeps the saw kerf open during ripping.

tively new design in the market. There are many jobs where this saw would be superior to a regular-cutting portable saw. How rapidly this saw is accepted into the market remains to be seen.

Accessories

Portable circular saws come equipped with some accessories; others are available as options in the market. They can be purchased from the original manufacturer or from a supplier. Some accessories can be used with any saw, while others are designed for a specific model.

When purchasing accessories for your portable circular saw, be sure they will fit or can be used with your saw. Read the specifications carefully. It is often difficult to return an accessory that does not fit or cannot be used with your saw.

Accessories can be divided into two groups: attachments and helpers. Attachments are those accessories which attach to your saw. Helpers are those accessories which help your saw by guiding it, containing it, or holding it.

Attachments The most common attachments used with the portable circular saw include the fence, splitter, dust collector and base or shoe lubricant.

Fence The fence is an attachment used to control the saw during rip cuts (Illus. 126). It usually clamps to the base (or shoe). It slides in a track for adjustment purposes. The distance from the fence to the blade determines the width of the rip (Illus. 127). Exact setup and ripping procedures are discussed in Chapter 5.

Some fences cannot be used on both the left and right sides of the blade (Illus. 128). Shorter fences can usually be mounted on either side of the blade. Unfortunately, a short fence does not offer as much control.

A new design offers a long fence that is reversible (Illus. 129). It is spring-loaded, and rotates on the control arm. You can position this fence on the left (Illus. 130) or the right (Illus. 131) of the saw by simply rotating it.

Splitter The splitter, which is used during rip cuts, is a piece of steel that attaches to the saw (Illus. 132). It is positioned behind the blade. As you make the rip cut, the splitter follows the blade into the wood, and keeps the saw kerf open as the cut progresses (Illus. 133). If the kerf begins to close, the splitter keeps the blade from being pinched. This minimizes the chance of a kickback.

The splitter comes with many European saws. It is not a common accessory on American-made portable circular saws. It is usually designed for use on a specific portable circular saw that uses a blade that cuts a kerf of the correct width. To be effective, the splitter must be slightly smaller than the width of the kerf. If the kerf is too wide for the splitter, the splitter will have little effect if the kerf closes on the saw blade and causes it to bind.

Dust Collector A dust collector, which cleans up the dust made by a portable circular saw, is an adaptor

Illus. 126A (above left). The fence controls the portable circular saw during rip cuts. It is positioned on the right side of the saw. Illus. 126B (above right). The same fence is positioned to the left of the saw in this photograph. Note how it attaches to the base.

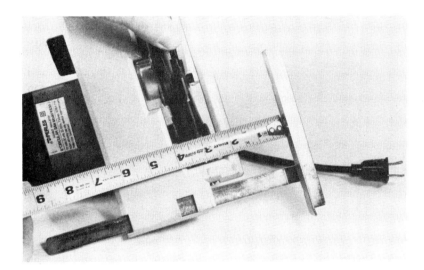

Illus. 127. The distance from the blade to the fence determines the width of the rip.

Illus. 128. This fence would be difficult to use on the left side of the saw due to its length and design.

Illus. 129. This reversible fence rotates on the control arm. This offers greater control during rip cuts on either side of the saw.

Illus. 130. The reversible fence has been mounted on the left side of the saw.

Illus. 131. The reversible fence has been mounted on the right side of the saw.

Illus. 132. The splitter attaches to the saw directly behind the blade. It follows the blade into the wood and keeps the kerf (saw cut) open.

Illus. 133. Note how the splitter rides in the saw kerf. The splitter keeps the kerf open even when stresses in the wood try to force it shut.

Illus. 134. This adapter fits over the sawdust chute and attaches to a common vacuum hose. Collecting dust at the sawing location makes the job neater.

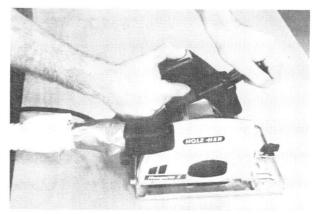

Illus. 135. The dust chute picks up the fine dust and particles. This device can be useful to remodellers who work in an occupied house.

become rough. This roughness could scratch sensitive materials such as plastic laminates or panelling. The that fits over the sawdust chute and connects to a common vacuum hose (Illus. 134). The dust collector is an attachment designed for a specific saw. When the dust generated by the portable circular saw is cleaned up, breathing conditions improve greatly. Some saws offer a dust bag similar to those on a belt sander. These bags contain most of the dust. If an adapter is used, a vacuum hose can be attached to the dust port.

The dust collector also reduces cleanup time when work is done on site, which is desirable to remodellers who are working in one or two rooms of an occupied house or business. The fine dust generated by the saw can travel throughout the building. Collecting the dust at the source reduces the mess and the cleanup problems (Illus. 135).

Base Lubricant The base or shoe lubricant is a self-adhering tape-like material that is applied to the base of the saw. It is designed to prevent scratching. With hard use, the base of the portable circular saw can

base lubricant makes the base smooth and reduces friction between the work and saw base.

Base lubricant is marketed under trade names such as Saw Slik™. Since the base configuration varies between saws, it is necessary to cut it to size. For best results, make a cardboard pattern for layout purposes. Cut the material with a utility knife or scissors.

Helpers The most common saw helpers used with the portable circular saw include the case, kerf splitters, saw guides, table- and radial-arm saw adapters, crosscutting jigs, and work supports.

Saw Case The saw case is an accessory that helps keep the saw clean, dry, and free of corrosion. Some saw cases are manufactured, and others are shop-made. Manufactured cases are usually metal or plastic, and shop-made cases are usually made of wood.

Most woodworkers prefer plastic or wood cases for their saws. Metal cases tend to rust and dent easily (Illus. 136). They do not withstand rough handling as well as wood or plastic. Plastic cases (Illus. 137A) usually have a moulded interior (Illus. 137B). They are usually designed for the shape of a specific saw, and are usually purchased from the saw's manufacturer. Some can be cut or modified to fit a different model of saw (Illus. 138), but it may be easier to build your own case.

The wood case shown in Illus. 139 is designed to contain most 7–8-inch portable circular saws. The interior is built to the shape of the saw (Illus. 140). The lid has a place to store extra blades (Illus. 141), and there is a pocket for wrenches (Illus. 142A).

Either buy or build a saw case. It protects the saw from the elements and prevents it from being damaged during transit. Plans for various saw cases appear in Chapter 11. The saw case is a good project to build with your new portable circular saw.

Kerf Splitters The kerf splitter is a saw helper that does the same job as the splitter on the saw (Illus. 143). It keeps the kerf open while the saw is cutting. This is how it works: Fit the splitter into the kerf. Tighten the wing nut to hold it in position. The kerf splitter will hold the two pieces in the same plane while you make the cut (Illus. 144). The most common kerf splitter is known as the KerfKeeper™.

Wedge-shaped stock can also be used to keep the kerf open. Shim stock or other pieces with a gradual taper can be driven into the kerf to keep it open (Illus. 145). Wedges will keep the kerf open, but they will not hold the pieces in the same plane. If your saw has a splitter attached, a kerf splitter may not be needed.

Saw Guide Saw guides are devices used to guide the path of the portable circular saw. They help the operator cut a straight line. Some saw guides clamp to the work. The saw rides along the edge of the guide (Illus. 146).

On two-part guides, one part (a straightedge) is mounted onto the work (Illus. 147–149), and the other is attached to the saw. The attachment fits into the straightedge and rides along its edge (Illus. 150 and 151). Some straightedges are designed for crosscutting and ripping. Others are designed only for crosscutting.

Some clamps are designed to be used as a saw guide (Illus. 152). They have no protrusions to interfere with the saw. Also, certain layout tools can be used to guide the path of the saw.

An inexpensive saw guide can be built in the shop or on the job. Mount the appropriate blade on the saw and measure the distance from the blade to the edge of the saw base. You can use either edge of the base. To this measurement add 2½ inches. Cut a piece of ¼-inch plywood this width. Note that the piece should be slightly longer (4–6 inches) than the piece you plan to cut.

Illus. 136. Metal cases are favored by some woodworkers; others feel that they rust and dent easily. They favor wood or plastic.

Illus. 137A (above left). Plastic saw cases come in many shapes. They are usually designed for a specific portable circular saw. Illus. 137B (above right). The insides of most plastic cases are moulded to accommodate a specific portable circular saw.

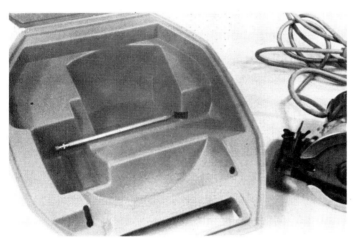 

Illus. 138A (above left). Moulded cases can be modified with a hand saw or utility knife to fit other saws. Illus. 138B (above right). This case was modified to fit another make of portable circular saw. Sometimes cases are sold at inexpensive prices, making it worthwhile to modify the case.

Illus. 139. This wood case was designed to contain most portable circular saws. Plans for this case are presented in Chapter 11.

Illus. 140. The saw is contained and held upright for travel. A piano hinge holds the lid securely in place.

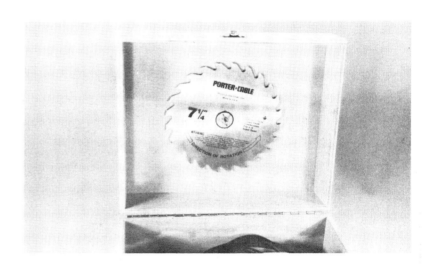

Illus. 141. The lid holds blades securely in a handy place. Remove the wing nut to release the blade.

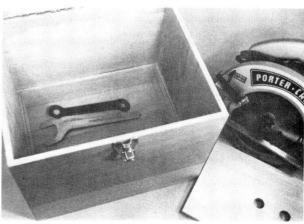

Illus. 142A (above left). The platform on which the saw rests is a container for wrenches and accessories. Illus. 142B (above right). Remove the lid to reveal the storage area. You may decide to build such a case for your portable circular saw.

Illus. 143. This kerf splitter is known as the Kerfkeeper™. Tighten the wing nut to hold it in position.

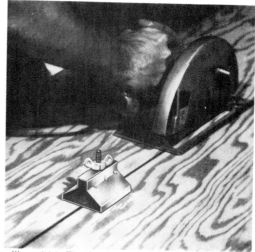

Illus. 144. The kerf splitter holds the kerf open and keeps both pieces in the same plane while you are cutting.

Illus. 146. This shop-made guide is clamped or nailed to the work. The saw rides along the edge of the guide.

Illus. 145. A wedge has been pushed into the kerf. The wedge will keep the kerf open as the cut progresses.

Illus. 147. From the underside, one part of this cut-off guide attaches to the saw and bolts to the other part of the guide.

Illus. 148. The fence rides along the edge of the work, and the saw moves to the left.

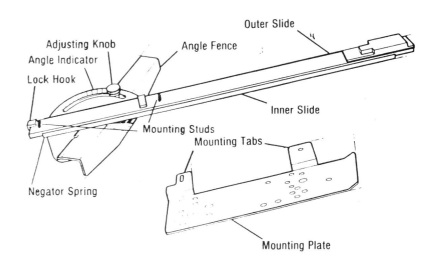

Illus. 149. This drawing shows how the parts of this cutting guide fit together.

Illus. 150. The fence of this cut-off jig is butted to the edge of the work. The saw blade is aligned with the layout line.

Illus. 151. The saw advances on the track to make the cut. Note that the stock has been clamped to the sawhorse for safer cutting.

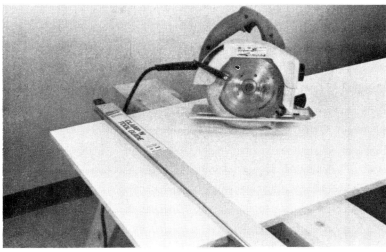

Illus. 152. This clamp is designed to be used as a saw guide. Note the clean design. It allows your saw to ride along the edge without obstruction.

Next, cut a piece of ½–1-inch plywood or particle board two inches wide. The plywood should have a true (factory) edge. It, too, should be slightly longer (4–6 inches) than the pieces you plan to cut (Illus. 153).

Glue and nail the ¼-inch plywood to the ½–1-inch plywood. The rough edge of the ½–1-inch plywood should be even with one edge of the ¼-inch plywood (Illus. 154). After the glue cures, clamp the guide to a sawhorse or other secure work surface. The ¼-inch plywood should extend over the work surface (Illus. 155).

Cut the plywood with the portable circular saw. The shoe of the saw will ride along the true edge of the ½–1-inch stock, and the blade will trim the ¼-inch plywood to the exact position of the blade's path (Illus. 156). Now, clamp the saw guide over the layout line. The blade will cut that path. There is no chance of making a measurement error when locating the saw guide (Illus. 157). *Note:* For best results, mount the blade you intend to use for general sawing to the saw before you cut the guide. This will make the guide more accurate.

Mitre Table- and Radial-Arm-Saw Adapters There are many saw helpers that are designed to transform a portable circular saw into a table saw or other stationary cutting guide. These devices usually have a cradle or mounting bracket into which the portable circular saw is mounted. The use of these devices is presented in Chapter 6.

Work Supports Work supports are devices on which the work is positioned for cutting. Sawhorses are the most common type of work supports. They can be either folding or rigid. Most folding sawhorses (Illus. 158) are commercially manufactured. Rigid sawhorses (Illus. 159) are usually shop-made. When selecting sawhorses, remember that their height and stability are important. The work should be at a comfortable cutting height.

Saw benches are also shop-made work supports. These benches are usually lower than sawhorses. Plans for saw benches shown in Illus. 160 and 161 can be found in Chapter 11.

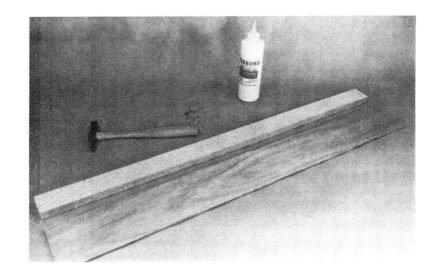

Illus. 153. These two parts can be combined to make a saw guide. Use parts slightly longer (4-6 inches) than the piece you plan to cut.

Illus. 154. Glue and nail the ¼-inch plywood to the ¾-inch stock. The best edge of the thicker stock should point inside.

Illus. 155. Clamp the cutting guide to the bench with the ¼-inch plywood hanging over the edge of the bench.

Illus. 156. The saw rides along the edge of the ¾-inch stock. The blade trims the ¼-inch stock to the proper width. Keep blade exposure low when making this cut.

Illus. 157. Locate this saw guide on the layout line; the blade will follow the line perfectly.

| A support for house painting. | The carpenter's best friend. | Portable picnic table base. | Temporary barricades. |
| Great storage for outboard motors. | Makes a base for a game table. | Handy support for sewing machine. | A base for study and hobbies. |

Illus. 158. Folding sawhorses are marketed commercially. They are easy to store and are quite rugged.

Illus. 159. These sawhorses are shop-made. They are quite rugged, but cannot be folded for storage. Plans for these sawhorses appear in Chapter 11.

Illus. 160. This saw bench supports long stock and makes a nice place to spend your lunch hour. Plans for this bench also appear in Chapter 11.

Illus. 161. This saw stool works well for supporting stock on the job. It also works well as a stool for installing trim around doors.

5

Basic Operations

There are a number of operations that the portable circular saw is capable of performing. These operations can be divided into three categories: basic, advanced, and specialty operations. The operations are classified according to frequency of use. Most novices will usually encounter the basic operations before the advanced or specialty ones.

Before attempting specialty or advanced operations, familiarize yourself with your saw. You can do this through the application of the basic operations, which include crosscutting, ripping, mitring, and cutting sheet and laminated stock. Be sure to also familiarize yourself with the general safety section of this chapter before attempting any operation.

General Safety Guidelines

Before operating any portable circular saw, you must read, know, and observe the following precautions. Review them periodically as you work. They will keep you alert to the common causes of mishaps.

1. Make sure that the portable circular saw is grounded properly. Use a ground-fault circuit interrupter when working in the field. Avoid working in damp places.
2. Make all adjustments and change the blade with the saw disconnected.
3. Make sure that the telescoping guard is working correctly. Never remove this guard or wedge it in the up position.
4. Set the blade exposure to no more than ¼ inch greater than stock thickness.
5. Make sure that the base is adjusted correctly and locked securely. Check the locking mechanisms periodically while working.
6. Make sure that the saw is equipped with the appropriate blade for the job.
7. Make sure that the stock is supported or clamped to a bench or sawhorse, and that the work does not pinch the blade as the cut is made. Thin stock must be supported close to the blade to eliminate blade binding.
8. Keep both hands on the saw when cutting. Never reach under the work or allow your hands to line up with the saw cut (in front or back of the saw).
9. Position yourself to the side of the saw. Avoid standing in the path of a kickback.
10. When using the rip fence, make sure that the pieces are at least 24 inches long. Shorter pieces may be pinched between the blade and fence. These pieces can be hurled with great force as the cut is completed.
11. Be sure to wear protective glasses when using the portable circular saw for any cut. The blade will throw chips in all directions. Eye protection is essential.
12. Set the base of the saw on the work before starting it. Make sure that the blade is clear of the work before starting the saw.
13. Allow the blade to come up to full speed before starting the cut. Keep the cord clear of the blade's path.
14. Know what is behind the work you are cutting. Make sure that you do not cut into electrical or plumbing supply lines. Cutting these lines could lead to an electrical shock.
15. Keep yourself balanced while cutting. Avoid overreaching. This can lead to a mishap.
16. As you complete the cut, make sure that the telescoping guard covers the blade before lifting the base off the work.
17. When cutting nonwood materials, make sure that the blades are compatible with your saw. Follow manufacturer's directions for safe operation.
18. When using portable circular saw accessories, make sure that they are compatible with your saw and that they are secured properly to the saw and work. Check the owner's manual for the saw and the accessory to be sure that the saw and accessory are compatible. When in doubt, do not use the accessory.
19. Keep the work area clean and free of scrap. Pick up cutoffs frequently.
20. Protect your respiratory system. When working with modified or treated wood, or with very fine wood dust, wear a dust mask.

Crosscutting

In general carpentry, the most common use of the portable circular saw is crosscutting. Crosscutting is done across the grain of the workpiece. In most cases, the carpenter marks a line on the workpiece, positions the saw, and makes the cut. The work is supported on a sawhorse or other work surface while the cut is being made. Many woodworkers use their knees to provide clamping pressure. This allows them to use both hands to control the saw. Never allow your knee to line up with the saw's kickback zone.

The steps involved in crosscutting are listed below. They will help you through the basic job.

1. Set up the saw. Check to be sure that the correct blade is mounted and that it is perpendicular to the base (Illus. 162). Do this with the power disconnected.
2. Adjust the blade exposure. Set the depth of cut to no more than ¼ inch greater than stock thickness. Do this with the power disconnected (Illus. 163).
3. Measure and mark the stock. Use a square to lay out the cutting line (Illus. 164).
4. Support the stock. Make sure that the stock is positioned on a sawhorse or other firm work surface (Illus. 165). *Never* use your leg or body to support work while it is being cut.
5. Make sure that the stock will not pinch the blade as the cut is being made. A short cutoff should **fall free**, and a long cutoff should be supported in such a way that it does not close the saw kerf on the blade.
6. Position the saw on the work. Make sure that the blade is on the waste side of the cutting line and is not touching the work (Illus. 166).
7. Turn on the saw and begin the cut. Hold the saw firmly with both hands. When the lower guard contacts the work, you will feel increased resistance (Illus. 167). Hold the saw firmly and continue cutting.

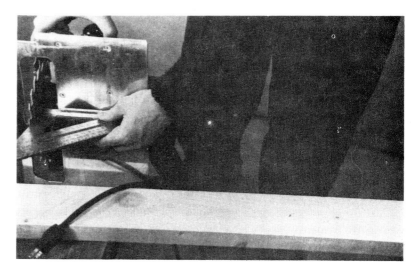

Illus. 162. When crosscutting, begin by checking the angle between the shoe and the blade. This angle should be 90 degrees.

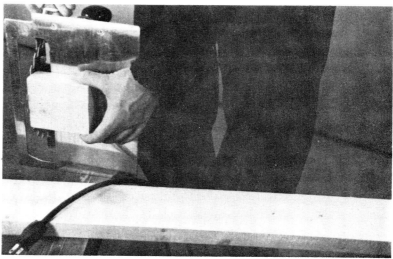

Illus. 163. Set blade depth so that it's slightly greater than stock thickness.

Illus. 164. Mark stock carefully. A bold line is easy to follow. Light lines tend to be obscured by sawdust.

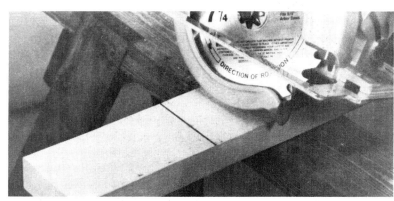

Illus. 165. Position your stock on a sawhorse or other work support. Work should be at a comfortable height.

Illus. 166. Position the saw on the work before turning it on. Be sure that the blade is on the waste side of the layout line.

Illus. 167. Hold the saw firmly with both hands. When the lower guard contacts the work, you will notice increased resistance. Push firmly to keep the blade on the layout line.

If you do not hold the saw firmly, the guard can move the saw off the layout line. Guide the saw completely through the work. Shut it off when you have completed the cut. The lower guard will return to its normal position. Do not put the saw down until the blade stops turning (Illus. 168).

8. If the saw binds in the kerf, shut off the saw. Binding usually occurs for one of two reasons: Either the blade is twisted in the kerf or it has become bound in the kerf. This happens because the cutoff bends towards the blade due to a lack of support. After you have determined what has caused the binding, repeat steps 6 and 7.

Crosscutting with the plunge-cutting saw follows a similar procedure (Illus. 169–172).

Some woodworkers use their squares or other guides to control the saw's path (Illus. 173–176). These guides are held with one hand, while the other hand controls the saw. In most cases, the guide will improve the cut; however, with practice and by using two hands to control the saw, you will be able to get square cuts without a guide (Illus. 177 and 178).

Check the end of your work with a square to determine its accuracy. This will help you identify any bad habits you may have. Practice cutting on some scraps before actually cutting the work, especially if you are using a saw with which you are not familiar.

Illus. 168. After the cut is completed, allow the blade to come to a complete stop before you release the saw.

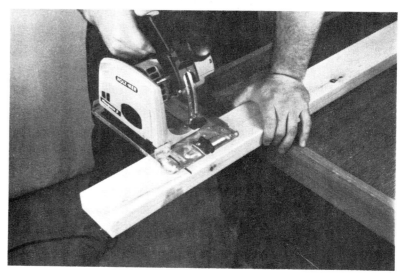

Illus. 169. When using the plunge-cutting portable circular saw, begin by positioning the saw on the work. The blade should be clear of the work and set for the correct exposure.

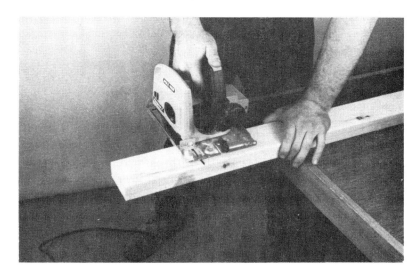

Illus. 170. Turn on the saw and release the plunge mechanism with your thumb.

Illus. 171. After the blade comes up to full speed and the plunge mechanism bottoms out, guide the saw into the work. Keep your eye on the layout line.

Illus. 172. When the cut is complete, release the plunge mechanism and shut off the saw. Let the blade come to a halt before releasing the saw.

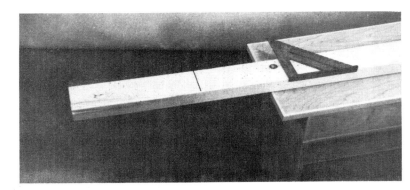

Illus. 173. A square or other guide can be used to mark the stock and guide the saw.

Illus. 174. Line up the saw with the cutting line. Make sure that the shoe of the saw is butted against the square.

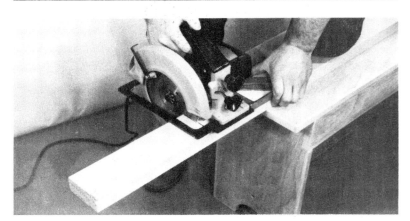

Illus. 175. The shoe or base of the saw rides along the edge of the square. Hold the saw firmly and use steady force to guide the saw.

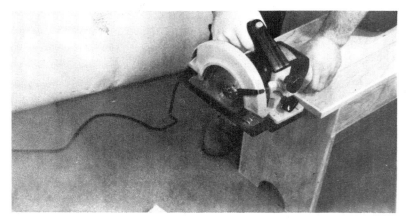

Illus. 176. The square keeps the saw blade from twisting as cutting progresses. Allow the lower guide to drop and the blade to stop before removing the saw from the work.

Illus. 177. This shop-made cutting jig also controls the path of the saw. It was made from the factory corner on a sheet of plywood.

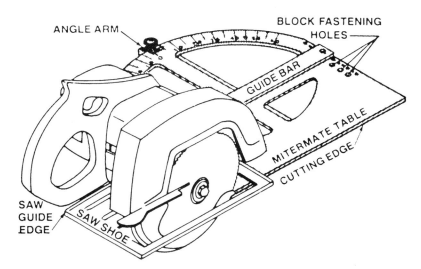

Illus. 178. This Mitermate™ Cutting Guide is a commercial jig that controls the path of the saw.

Ripping

Ripping, a cut made with the grain, reduces the width of the workpiece. This cut requires more energy than crosscutting because of the orientation of the wood fibres and the blade's tooth shape. Rip blades are generally coarser than crosscut blades, which means that the blade takes a larger bite and throws a heavier chip. More energy is needed to cut and eject these chips.

For rip cuts, mark the stock with a pencil line or clamp a rip fence to the base on the portable circular saw. The stock must be supported and held stationary. The steps involved in ripping are listed below. They will help you through the basic job.

1. Set up the saw. Check to be sure that the correct blade is mounted and that it is perpendicular to the base. Do this with the power disconnected. A saw equipped with a splitter can make ripping safer and more efficient.

2. Adjust the blade exposure. Set the depth of cut to no more than ¼ inch greater than stock thickness. Do this with the power disconnected.

3. Measure and mark your stock. Use a combination square to mark a line parallel to the edge (Illus. 179), and set the fence to the desired width. Use a tooth that points towards the fence to make this setting. You may want to test the setup on scrap to make sure that it is correct. Remember that the blade must be positioned in the waste stock (Illus. 180).

4. Support and clamp the stock. Make sure that it is positioned on a sawhorse or other firm work surface, and that the blade does not cut the supports. Stock must be clamped because the thrust of the saw is directly into the work. It is *not* safe to hold the workpiece with one hand while ripping with the other.

Illus. 179 (above left). Lay out the rip with a combination square or other layout tool. Long rips on irregular stock can be laid out with a chalk line. Illus. 180 (above right). If you are using a rip guide, set it to the desired width. Make sure that the blade is on the waste side of the layout line.

Your hands are always on a collision course, and I would never want to collide with a rip blade!

5. Have kerf splitters or wedges available to keep the kerf open as the rip cut progresses through the workpiece. A splitter on the saw will usually keep the kerf open, but kerf splitters or wedges can be used in conjunction with a saw equipped with a splitter.

Remember, binding in the kerf will tax the blade and motor, and is also a potential cause of kickback. *Never* position your hands behind a portable circular saw. In the event of a kickback, the blade could cut your hand. It is very difficult to cut yourself when both hands are on the saw.

6. Position the saw on the work (Illus. 181). Make sure that the blade is on the waste side of the cutting line and is not touching the work.

7. Turn on the saw and begin the cut (Illus. 182). Hold the saw firmly with both hands. When the lower guard contacts the work, you will feel increased resistance. Hold the saw firmly and continue cutting. If you do not hold the saw firmly, the guard can move the saw away from the layout line.

Illus. 182. Turn on the saw and guide the blade into the workpiece. Hold the saw firmly. Use both hands to guide the saw. Note that the work has been clamped to the sawhorse.

8. As you progress through the work, place kerf splitters or wedges in the kerf to keep it open (Illus. 183). Shut off the saw when placing splitters or wedges.

Illus. 181. Position the saw on the work. Make sure that the blade is lined up correctly with the cutting line.

Illus. 183. The splitter on this saw follows the blade into the saw kerf. Use a wedge to keep the kerf open if your saw is not equipped with a splitter.

Illus. 185. A saw with a large base would be difficult to control on this small piece.

9. Guide the saw completely through the work. Shut off the saw when the cut is complete. The lower guard will return to its normal position. Do not put the saw down until the blade stops turning (Illus. 184).
10. If the saw binds in the kerf, shut off the saw. Binding usually occurs for one of two reasons: Either the blade is twisted in the kerf or it has become bound in the kerf. When binding occurs, it is usually caused because not enough kerf splitters or wedges have been used. A blade with too little set can also cause kerf-binding. After you have determined the cause of blade-binding, repeat steps 6, 7, and 8.

Illus. 186. A smaller saw would be easier to control on this rip cut. Note that the stock has been clamped to control the work.

Illus. 184. Allow the blade to stop turning and the lower guard to drop before you release the saw.

Ripping Small Pieces When ripping small pieces, select a saw that fits the job (Illus. 185 and 186). A large saw is awkward on a small job. Clamp the work securely to a work support. Make sure that the blade has clearance and does not cut the work support (Illus. 187). Work only as fast as the blade will cut. Adjust clamps to clear the work, as needed. Add a wedge to the kerf to keep it open for the rest of the cut (Illus. 188). Continue the cut. Be sure that the blade is clear of the wood as you restart the saw (Illus. 189). Complete the rip cut (Illus. 190). Release the saw only after the blade comes to a stop.

Heavy Rips A heavy rip is a rip in stock 1½ inches thick or thicker. These cuts require a great deal of energy. There is more blade in the stock, and a coarser blade may be needed. Coarser blades cut faster and do not cause the portable circular-saw-motor to overheat. Fine blades can cause overheating due to slow feed.

Though heavy ripping can be done in the field with the portable circular saw, use a stationary circular saw if you have to do a great deal of heavy ripping. Heavy

Illus. 187. The saw blade cuts into the rip slot in this sawhorse. This eliminates damage to the work support.

Illus. 188. Reposition the work and clamp it again. Insert a wedge in the saw kerf to keep it open.

Illus. 189. Resume cutting; keep the blade clear of the wood until the blade comes up to full speed.

Illus. 190. Release the saw only after the blade comes to a complete stop.

ripping is slow. It consumes a great deal of labor and can burn out a circular saw motor quite easily.

Worm-drive circular saws are usually best suited for continuous heavy ripping (Illus. 191–197). But there are many direct-drive saws that will work as well.

Heavy rips are usually made in wood with a relatively high moisture content. This means that the wood chips weigh more and more energy is required to eject them. Avoid using a light- or a medium-duty portable circular saw for heavy ripping. This is the fastest way to burn up a portable circular saw.

Work carefully during a heavy rip. With most of the blade engaged in the wood, a kickback is much more likely. Twisting or turning the saw even slightly could cause a kickback. Keep both hands on the saw, and make sure no part of your body is in the kickback zone.

If the blade is not large enough to cut through the work, make a second cut on the opposite face. This kind of cutting is known as resawing, and requires accurate layout and good control of the saw (Illus. 198–202). For best results, make the first cut as deep as possible. If a deep cut strains the saw, however, then take a lighter cut. The blade will follow the path of least resistance on the second cut. It will follow the kerf made with the first cut.

Be careful how you clamp the work. If the blade is pinched during the second cut, it is likely to kick back.

Some larger saws are capable of making heavy rips in a single pass (Illus. 203 and 204). These saws are very powerful, and should be handled carefully. While most of these saws are equipped with brakes, they are capable of kicking back during a cut.

Avoiding Kickbacks When you are making rip cuts, kickbacks are *always* a possibility. Certain factors contribute to kickbacks. To avoid these factors, remember the following: Saw blades should be free of pitch, have an adequate set (to keep from binding), run true (free of twist of warp), and have minimum exposure (¼ inch greater than stock thickness). Stock should be dry, free of knots, not warped or twisted, supported firmly (to keep blade from being pinched), and clamped securely.

Illus. 191. Heavy rips are best done with a worm-drive saw or heavy-duty direct-drive saws. Note that the work has been clamped to the sawhorse.

Illus. 192. To avoid cutting the sawhorse, stop here and move the sawhorse.

Illus. 193. Continue making the rip cut after you have shifted the sawhorse and clamped it to the work.

Illus. 194. Insert a wedge in the saw kerf to minimize the chance of a kickback. Stand to the side of the work and keep both hands on the saw.

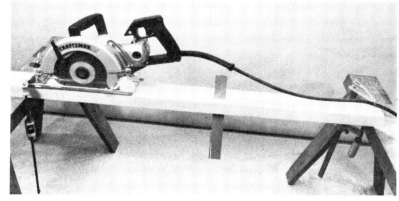

Illus. 195. As the rip progresses, move the other sawhorse.

Illus. 196. Resume cutting after repositioning the sawhorse.

Illus. 197. Complete the rip cut using the appropriate ripping practices.

Illus. 198. When resawing (ripping into thinner pieces), set the blade to full depth for the first cut. If a deep cut strains the saw, then take a lighter cut.

Illus. 199. Use the standard ripping practices. Keep the kerf open and stand to the side of the work. Follow the layout line carefully.

Illus. 200 (above left). Keep the stock clamped while ripping. Control of the workpiece makes ripping safer and more accurate. Illus. 201 (above right). Allow the saw to clear the stock completely before lifting the saw. Flip the stock over and cut the work from the opposite edge.

Illus. 202. The work has been resawed. Note that the cutoff is uneven. The layout line on the work was followed closely so that the workpiece is truer.

Illus. 203. This 10-inch saw will rip the work in a single cut. Heavy rips should be performed with the utmost care.

Illus. 204. This 16-inch saw will rip stock much thicker than the work shown. Keep blade exposure to a minimum.

Mitre Cuts

Mitre cuts are angular cuts made across the face, end, or edge of the work. Most mitre cuts are made at a 45-degree angle so that when two pieces are joined they make a 90-degree angle. This is common in door trim and window frames. Mitres can also be made at other angles for specialty work such as rafter ends, stair stringers, and trim.

Mitre cuts require accurate layout and careful saw setup. Test the setup in scrap before actually cutting the workpiece.

Face Mitres A face mitre is cut across the face of the work. It is similar to a crosscut. The blade is set perpendicular to the base for this cut. You can cut face mitres freehand along a layout line or use a guide for the cut (Illus. 205 and 206). A cutoff guide controls the path of the saw blade to ensure accuracy. It uses the edge of the work for control (Illus. 207–211). An adjustable guide, straightedge, or cutting guide can also be used for this purpose. Clamp or nail the straightedge or cutting guide to the work.

Make sure that you start the saw correctly when beginning a face mitre. If you start the cut incorrectly, the mitre will not fit well against the mating part. Push the saw into the work at a uniform speed. Avoid stopping during the cut; this usually causes the blade to veer from the intended course.

Mitring attachments for cutting face mitres with the portable circular saw can also be used to increase the accuracy of the cut. These attachments are discussed in Chapter 6.

End Mitres End mitres are also a crosscut. Set the blade at the desired mitre angle before making the cut. End mitres are usually cut at a 90-degree angle to the edge of the work. A crosscut guide can be used to control the cut (Illus. 212–214), but it can also be cut freehand.

If you are cutting an end mitre in the middle of a piece, make sure that the work is supported well, or it may pinch the blade. This could cause a kickback. It can also affect the accuracy of the cut and leave burn marks on the cut surfaces. Remember to limit blade exposure to slightly more than what is needed to separate the piece.

Illus. 205. A commercial guide can be used to improve the quality of your mitres or compound mitres.

Illus. 206. This commercial guide can be used to cut flat or compound mitres on stock up to approximately 8 inches wide.

Illus. 207 (above left). This commercial guide was set to the desired mitre angle and clamped securely. Illus. 208 (above right). Line the blade up with the cutting line and butt the guide against the base or shoe of the saw.

Illus. 209. Turn on the saw. Begin cutting when the saw comes up to full speed.

Illus. 210. Push the saw at a uniform speed. Slowing or stopping the saw can cause it to veer off the cutting line.

Illus. 211. The completed mitre is smooth and even, the result of good sawing practices.

Illus. 212. Set the portable circular saw to the mitre angle. The cut-off guide will control the path of the saw.

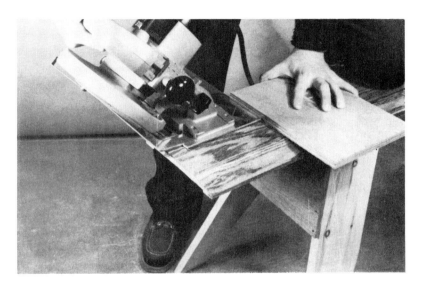

Illus. 213. Position the saw on the work and adjust the blade to the cutting line.

Illus. 214. The results: This end mitre shows the quality that can be achieved with a portable circular saw.

Edge Mitre The edge mitre is a rip cut with the blade set to the desired mitre angle. Edge mitres can be made freehand or with a guide. For absolute accuracy, use a fence (Illus. 215–217).

Make sure that the work is held securely before you begin. Set the blade depth to slightly greater than stock thickness. For best results, use a saw that has a splitter. If none is available, use wedges to keep the pieces separated during the cut.

Turn on the saw and guide it into the work. Start carefully to ensure an accurate cut. Feed only as fast as the machine will cut. Remember, you have more blade in the work because of the blade tilt. Ripping also requires more energy than crosscutting, so the feed may be slower. If the slow feed causes burning, use a coarser blade.

Use a wedge to keep the kerf open as the cut is being made. Keep an eye on the layout line as you freehand cut. If you are using a fence, look at it occasionally to be sure that it remains in contact with the edge of the work (Illus. 218).

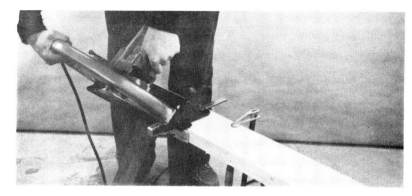

Illus. 215. This edge mitre is an angular rip cut. A fence is being used to control the saw.

Illus. 216. Note that the stock was clamped to the sawhorse while it was being cut. Use standard ripping practices to make this cut.

Illus. 217. When the saw is tilted, the center of gravity changes. This changes the feel of the saw. Maintain firm control while making the cut.

Illus. 218. This edge-mitre cut will form the angular edges of a sawhorse. Complete plans for the sawhorse appear in Chapter 11. Note that the fence rides smoothly along the edge of the work.

Cutting a Taper

Taper cuts, which are made on legs, braces, and wedges, are relatively common cuts to make with the portable circular saw. It is usually best to begin a taper cut at the widest end of the work (Illus. 219–221); this keeps the lower guard from catching on the thin pointed part of the work.

Clamp or anchor the work securely. In some cases, the taper will seem like a crosscut; in other cases, it will seem like a rip cut (Illus. 222). Handle the saw accordingly. Taper cuts can be made freehand (Illus. 223) or with a guide. Keep the saw against the guide or travelling along the line.

Avoid forcing the saw into the work; this can cause you to leave the line or guide. Keep blade exposure to a minimum for the most accurate sawing. It may be necessary to lift the guard as the cut begins. This will keep it from catching on the workpiece or travelling off course.

Compound-Mitring

Plywood roofing parts and framing braces often require compound-mitre cuts (Illus. 224). A compound mitre differs from a conventional mitre. The conventional mitre has one angular cut. A compound mitre is an angular cut with the blade tilted. There are actually two angles to be concerned with (Illus. 225).

Prepare to make the compound mitre by setting the blade at the desired angle. Lock the setting securely. Lay out the work with the desired angular cut and set the blade exposure to cut through the workpiece (Illus. 226). You can control the saw with a cutoff guide or straightedge, or make it freehand.

Make sure that the work is supported and held securely before cutting. Allow the saw to come up to full speed before you start the cut (Illus. 227). Guide the saw carefully into the work. If you are using a guide, check the work to be sure that the cut is following the layout lines. Feed the saw through the work (Illus. 228 and 229). Make sure that the work remains supported as the cut is being made. Feed the saw completely through the work before shutting it off (Illus. 230).

If the saw binds during the cut, the blade is probably pinched between the two pieces. This can occur due to the pressure of unsupported or unclamped pieces binding against the sides of the saw blade. Avoid this situation, as it is hard on the saw and may cause a kickback. It is also very difficult to get an accurate cut if the blade becomes pinched.

Illus. 219. This taper-cutting setup allows little control of the saw. Note that the base or shoe is not touching the workpiece.

Illus. 220. A left-hand saw would improve this setup. Now the base of the saw rides on the workpiece.

Illus. 221. The workpiece was marked on the opposite side so that the right-hand saw could be used correctly.

Illus. 222. For taper-sawing, it is best to begin on the widest end. The guard may have to be lifted when the cut is started.

Illus. 223. For best results, concentrate on the layout line and the path of the saw. When stock is clamped, it is easier to concentrate on sawing. There is little chance of stock movement.

Illus. 224 (above left). Roofing jobs such as this often require compound-mitre cuts on the rafters and the plywood sheathing. *Illus. 225 (above right).* When making a compound-mitre cut, there are actually two angles to be concerned with. When the blade is set at an angle, there is more blade in the wood. This may require a slower feed.

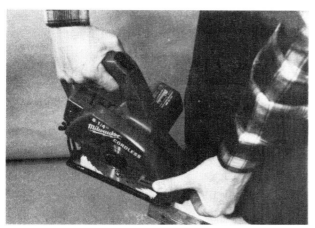 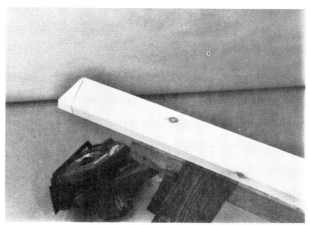

Illus. 226 (above left). When making compound mitres, first set the blade exposure. It should be about ¼ inch greater than stock thickness. *Illus. 227 (above right).* For best results, lay out the compound mitre carefully. This is even more important when sawing freehand.

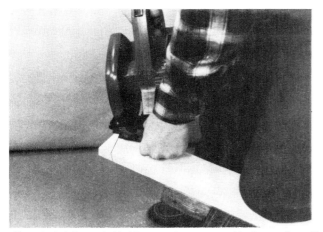 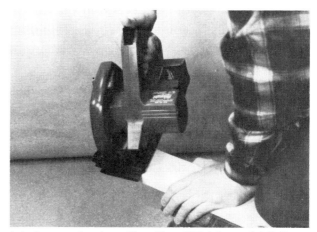

Illus. 228 (above left). Line up the saw and balance it on the work before you begin cutting. *Illus. 229 (above right).* Turn on the saw and feed the blade into the wood at a uniform speed.

Illus. 230. When the cut is complete, shut off the saw and let the blade come to a stop before you release the saw.

Cutting Plywood and Sheet Stock

Plywood and other forms of sheet stock are frequently cut with a portable circular saw. Sheet stock is used for floor underlayment, panelling, roofing, and siding jobs. Most of these jobs take place on site where stationary equipment is not available (Illus. 231–241).

Sheet stock must be supported well to keep the pieces from binding against the saw blade. Usually, two-by-four stock is used to lift the piece above the pile and support it during the cut (Illus. 242).

Mark the stock carefully and accurately (Illus. 243–245). For finish work, use a guide to control the saw; this improves the quality of the cut. Select the appropriate blade for the job. Usually, a fine-cutting carbide-tipped blade works best. Tool-steel blades dull quickly when you cut particle board or other composition materials.

On sensitive materials such as plywood and panelling, the best cut is made with the good face of the stock down (Illus. 246–252) so that any grain tear-out will be on the back or poorer face of the work. Generally, a good, sharp carbide-tipped blade will not tear the grain if it travels on a straight line and does not flutter. Blade flutter occurs when there is runout in the arbor, which occurs on saws where the motor or arbor bushing has been worn through use. Saws with ball or roller bearings around the arbor are not likely to flutter. The bearings keep the arbor turning in a true orbit.

Set the depth of cut slightly deeper than the stock thickness. Move the supports close to the work, and make sure that the supports are not in the blade's path. You may have to secure the stock with clamps to keep it from moving while the cut is being made, especially if working with thin stock. The thrust of the saw will actually move the panel.

Turn on the saw and let it come up to full speed. Feed the blade into the work. Keep the saw against the guide or straightedge if you are using one. Make sure that the cord is clear of any interference. If the cord gets caught, it could pull the saw off course.

If you are cutting freehand, try not to turn the saw. This is likely to cause tear-out or splintering. Guide the saw completely through the work so that the lower guard will spring back into place. Turn off the saw.

Jobs such as underlayment for vinyl floor covering and wall panelling may require several cuts in a piece of sheet stock. It is smart to number the cuts if there are several (Illus. 253). This practice provides extra support for any thin parts. It also saves time and ensures that you only cut away the waste stock (Illus. 254).

Most underlayment jobs begin in a corner. Make sure that the corner is square before you lay out the sheet. For complicated cuts, it may be helpful to make a cardboard pattern. This will eliminate errors or waste (Illus. 255).

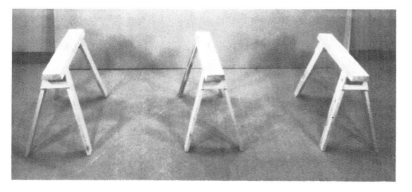

Illus. 231. These sawhorses are used for cutting sheet stock in the field.

Illus. 232. The fold-out work support is used with the sawhorses to support sheet stock.

Illus. 233. As the edge of the support is lifted, it turns into a grid.

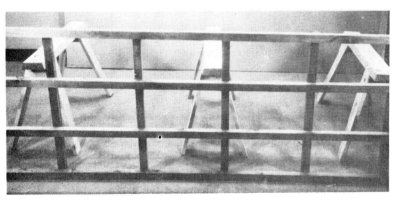

Illus. 234. The grid can be used with the long members up for ripping. Put the short members up for crosscutting.

Illus. 235. Position the work support for ripping.

Illus. 236. Position the particle board on the work supports.

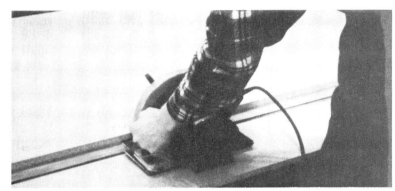

Illus. 237. Clamp a guide to the work; the saw rides along the guide.

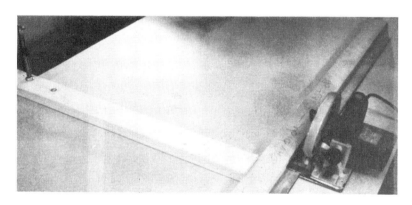

Illus. 238. Clamp a strut to the work. It supports the cutting guide. This guide could deflect without the strut.

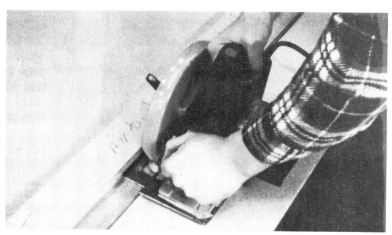

Illus. 239. The saw is controlled by the guide. Keep both hands on the saw for best results.

 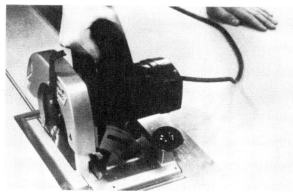

Illus. 240 (above left). A commercial saw guide can also be used to control the saw. Illus. 241 (above right). The shoe of the saw rides along the edge of this commercial saw guide. Observe common sawing practices when using any saw guide.

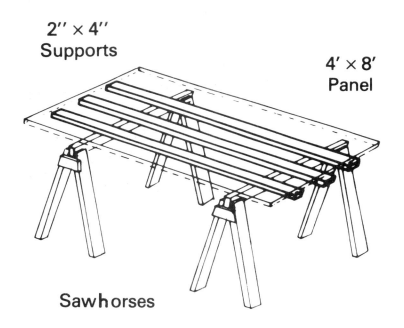

Illus. 242. Large panels should be supported on framing lumber. This gives the panels additional support while cutting. Make sure that the blade exposure is minimal. This will eliminate the chances of cutting the framing stock in half.

Illus. 243. After marking the stock, use a straightedge to guide the saw through the cut. Less tear-out occurs when panelling is cut with the good side down.

 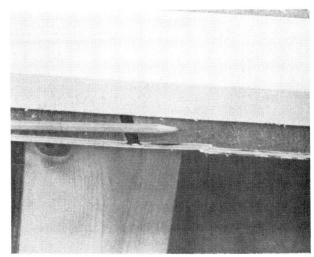

Illus. 244 (above left). No tear-out occurred during this cut. A high-quality blade reduces the chance of tear-out. Usually, a high alternate-top-bevel-tooth configuration is preferred. Illus. 245 (above right). The base of the saw slid under the saw guide on this cut. The trim moulding will hide this error, but it is a problem when you are cutting thin panelling.

Illus. 246. Position spacers on the workbench. They lift the work off the bench and allow clearance for the blade.

Illus. 247. Lay the panel good face down on the spacers.

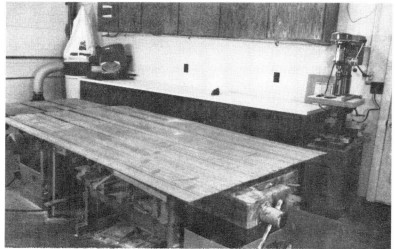

Illus. 248. Lay the panel out carefully. Remember, the cuts are backward when the panel is upside down.

Illus. 249. Clamp the straightedge guide along the layout line. Then move the spacers in near the cut. One supports each side of the cut.

Illus. 250. Begin cutting with the base of the saw butted to the cutting guide.

Illus. 251. Position the work at a comfortable height so that you do not have to overreach.

95

Illus. 252. Tear-out is on the back side of the panel when it is cut this way. A finer blade could reduce tear-out.

Illus. 253. If there are several cuts, number them.

Illus. 254. Cut away only the waste stock.

Illus. 255. For complicated cuts, it is helpful to make a cardboard pattern.

Cutting Hardwood Plywood and Laminated Sheet Stock

Today, most hardwood plywood has a very thin veneer on the face and back that tends to splinter when cut. For the best cutting, cut the work with the exposed face down. Grain tear-out will occur on the back side as the blade leaves the work.

Blade selection can minimize the problem of tear-out in plywood cuts. In most cases, an alternate top bevel or triple-chip tooth will generate less stress and tear-out. Flat-top teeth or tool-steel blades with a large set can cause severe tear-out.

It is also possible to score the wood along the cutoff line; this makes the wood fibres along that line the weakest. As the fibres begin to tear, they break off evenly at the scored line. Some cutting guides also help minimize tear-out because the base of the guide reinforces the wood fibres at the cutting line.

Some woodworkers use masking tape to eliminate tear-out. The masking tape reinforces the fibres and makes them more resistant to tear-out. The tape is applied over the layout line, and the saw blade cuts through it. Use a great deal of care when removing the tape because it may actually lift the veneers from the core.

Sheet stock laminated with melamine or other plastic laminates may also be a problem (Illus. 256–258). In an ideal work situation, they should also be cut with the good side down, but this is not usually possible. Flipping the work over can actually damage the laminate worse than the blade would. In addition, many laminated counters require cutouts after installation. These cuts are made on site with the good side up.

Blade selection is important here. A 40-tooth carbide-tipped blade with a triple-chip or alternate top-bevel grind (7¼-inch blade) would work best. If the saw is of high quality and the operator is skilled, a smooth cut can be made. Saws with arbor runout tend to cause tear-out. Turning or twisting the saw in the cut may also cause tear-out.

When working with hardwood plywood, panelling, or laminated sheet stock, it is important that you use a saw with a smooth base. A scratched saw base can do more damage to the work than the blade. Select a saw with a good base for finished cuts. Use a base coating if necessary. Damage to finished work can be expensive.

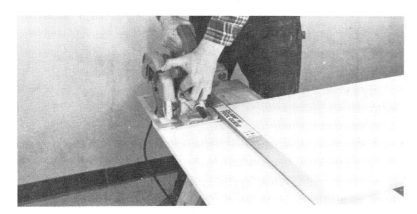

Illus. 256. This laminated particle board (Kortron™) is cut the same way any other sheet stock is cut. A guide helps to minimize the possibility of tear-out.

Illus. 257. Do not force the saw when making this cut. Pushing the cut can cause increased tear-out.

Illus. 258. This completed cut shows no tear-out. The correct blade and a saw guide help eliminate the problem. Smooth cuts like this are dependent on the saw blade, a saw guide, and the skill of the woodworker. The more cutting you do, the more skilled you become.

Getting a Straight Cut

Though experience is probably the biggest factor in getting a straight cut with a portable circular saw, there are many other factors. These include as follows: a guide, proper blade selection, blade condition and exposure, saw condition, feed speed, and hardness of the material.

Setting up a guide is time-consuming, but for many cuts it is time well spent. The saw is much easier to control, and the results are much better. A guide is no guarantee of a straight cut, however. The shoe or base of the saw can also act as a guide (Illus. 259 and 260). Clamping the work will also make getting a straight cut easier (Illus. 261).

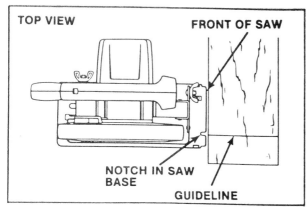

Illus. 259. The notch in the base or shoe can act as a guide. Line it up with the cutting line.

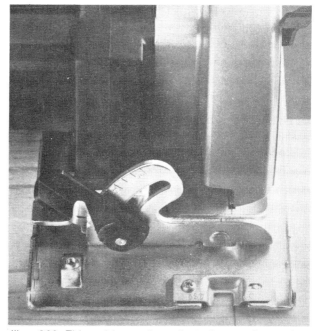

Illus. 260. This guide can be adjusted according to the blade you use.

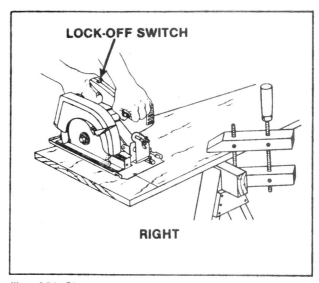

Illus. 261. Clamping the work helps you get a straighter cut because two hands can be used to control the saw and you are not distracted by the movement of the workpiece.

The condition of the saw and blade are very important. A dull blade may actually cause the saw to veer off course. If the teeth are dull on only one side of the blade, the cut will travel in the direction of the sharper teeth. Blade flutter from a thin blade or arbor runout may also cause the saw to veer off course.

Keep blade exposure low. Too much blade in the work will make it difficult to turn the saw slightly while cutting. It is also much more dangerous.

A fine blade will require a slower feed than a coarse one. Forcing a fine blade into the work can cause it to turn because the saw blade is loaded with chips and the additional force causes it to turn. Feed the blade into the work only as fast as it will cut without stalling or slowing down.

When a saw stalls or slows down due to overfeed or hardness of the material, the blade has a tendency to run off course. An electronic saw may be helpful here, since it senses changes in blade speed and responds with increased energy.

Portable circular saws with a splitter may also be easier to control once the cut is established because the splitter rides in the kerf and controls the saw's travel. If the cut starts straight, the cut remains straight, but the opposite is also true. For rip cuts, the splitter is a great help for minimizing the chances of kickbacks.

6
Advanced and Specialized Operations

As you become more skilled with your portable circular saw, you will want to perform some of the following advanced operations. Be sure to reread the safety precautions found in Chapters 2 and 4 before attempting any operation. Specific safety practices will also be discussed with each operation listed.

Plunge-Cutting

In many cases, a blind cut or rectangular hole must be made with the portable circular saw. This type of cut is called a plunge cut, and is made for electrical outlets, sink cutouts, and plumbing and heating access holes. Plunge cuts can be made with a conventional portable circular saw or a plunge-type portable circular saw. These techniques are discussed below.

Regardless of which type of plunge cut you make, preparation is essential. Be sure to lay out and mark the cut carefully. Determine whether the blade must be on the inside or outside of the layout line. Check the underside of the work for obstructions such as pipes, wiring, framing timbers, nails, and supports such as sawhorses. Cutting through such objects could cause electrical shock and damage to the saw or blade.

Conventional Plunge-Cutting The conventional saw can be used for a plunge cut. The first step is to set the blade exposure to a distance slightly greater than the thickness of the work. Make sure that the work is properly secured when working on sawhorses or a bench.

Begin in any corner. Lift the guard and eyeball the position of the blade with the corner of the cut. Make sure that the blade is parallel with the layout line. The front of the saw base must be resting firmly on the work, and the rear of the base must be high enough to allow the blade to clear the work.

Grasp the saw firmly with both hands. Hold the guard up with your thumb. Turn on the saw and allow it to come up to full speed. Slowly lower the blade into the work.

Illus. 262. Begin a plunge cut with the lower guard retracted. Once the blade has come up to full speed, slowly lower it into the work.

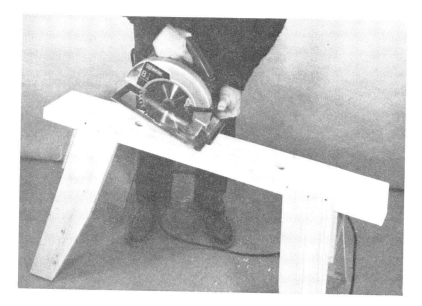

Illus. 263. Keep the blade on the waste side of the layout line. The saw will cut in the normal fashion once the plunge is complete. Never back up when making a plunge cut.

Illus. 264. Cut out the ends with a hand saw. The cuts should meet correctly if the plunging action was exact.

Keep an eye on the layout line to be sure that the blade remains parallel to it (Illus. 262–264). When the base of the saw is firmly on the work, push the saw forward and begin cutting in the normal manner. When you reach the end of the cut, shut off the saw. Allow the blade to completely stop before you retract the saw from the work.

When plunge-cutting, do not back up with the saw. This could cause a kickback. If you miss the corner on the first cut, make another plunge cut closer to the corner, but avoid backing up. When a complete rectangle must be removed, a series of four plunge cuts is needed. If the stock is heavier than ¼ inch, it may pinch the blade when the fourth cut is made.

On heavier stock, you can screw a sheet-metal screw into each saw kerf as cutting progresses. The screws hold the cutout in position when the fourth cut is being made. To remove the cutout, simply remove the screws. This procedure eliminates binding on the saw blade or possible breakage before the cut is complete.

When you are cutting thinner stock, binding is not usually a problem. Make sure that the stock is supported firmly around the cutout area. Thin stock has a tendency to move around during the cut because of its light weight, and it is important that you clamp it. Do not try to hold it with your hand. Both hands are needed to control the saw.

With practice, you will find plunge-cutting to be as accurate as any other method, and faster than many other methods. Work carefully, and practice on scrap before actually cutting the work.

Using a Plunge-Cutting Saw The plunge-cutting saw has no lower guard to get in the way. It has a spring-loaded base. When you push the motor portion downwards, the blade drops down through the base (Illus. 265). You have to remove the splitter for plunge-cutting. Set the blade exposure to no more than ¼ inch greater than stock thickness. Position the saw near the corner of the cutout area. You will have to eyeball the position of the saw (Illus. 266).

Rest the base of the saw firmly on the work. Note that the hole in the base can be aligned with the layout line; this provides greater accuracy when you are making this cut. Turn on the saw, allow it to come

Illus. 265 (above left). This plunge-cutting saw can plunge from a fixed position with no blade exposure. Illus. 266 (above right). Line up the saw with the cutout and turn the saw on. The plunging action is straight down.

up to full speed, and slowly push down on the motor. Both hands should be on the saw. When the motor bottoms out on the stop, begin cutting in the normal fashion.

When you reach the end of the cut, shut off the saw and retract the blade. The saw can now be lifted off the work (Illus. 267). If you find that the blade has not cut completely to the corner, make another plunge cut closer to the corner. Do not back up. This can cause a kickback.

If you are removing a complete rectangle from the work, follow the practices discussed under conventional plunge-cutting.

Illus. 267. The plunge cut is efficient and safe. Both hands are on the saw, and the blade is always guarded.

Cutting Irregular Shapes

Irregular shapes are a challenge to cut because the base may have to "ski" over irregularities or ride tangent to an arc. In past experiences, I have used the portable circular saw to cut such irregular shapes as logs, church pews (Illus. 268–272), and post-formed counter tops.

Irregular shapes must be clamped or held securely for cutting (Illus. 273). In some cases, specialty guides can be built to control the saw's path. It may be difficult to clamp guides to an irregular shape; but you may be able to nail them to the work. Select the nailing position carefully. This will make the nail holes less obvious when the cutting is complete.

Blade exposure may have to be greater than normal on irregular surfaces. This is because stock thickness may vary on the work. This was the case with the church pews. The stock was much thicker at the front edge of the pew.

Use a great deal of caution when a large amount of blade is exposed. It is more difficult to get a straight cut when a large amount of blade is in the wood. There is also a greater chance of kickback.

The base of most portable circular saws is upturned at the front. This allows it to "ski" over rough or irregular stock. Once you get the feel of any saw, you can help it ski over irregularities. You can use the hand you place on the control knob to actually lift the saw, which makes it easier for the base to ski over the surface. On some irregular surfaces, the blade may actually bind or kick back as the saw base changes planes. Work with care and be aware of the kickback potential.

Illus. 268. This church pew was clamped to the sawhorses; one end of the saw guide was clamped to the pew.

Illus. 269. The other end of the guard was nailed to the pew. The nail was driven in a position where the hole would not be obvious.

Illus. 270. The saw rode along the guide and the cut was made in the typical fashion. Increased blade exposure was required because of the irregular shape.

Illus. 271. Next, the guide was nailed to the bottom of the church pew. Cutting proceeded from the bottom of the pew.

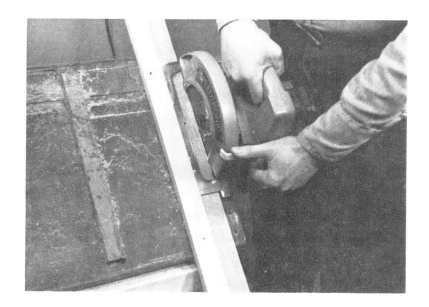

Illus. 272. Careful layout, a cutting guide and firm clamping produce excellent results on an irregular surface.

Illus. 273. Irregular shapes must be held securely for sawing. Clamps or nails can be used for this purpose.

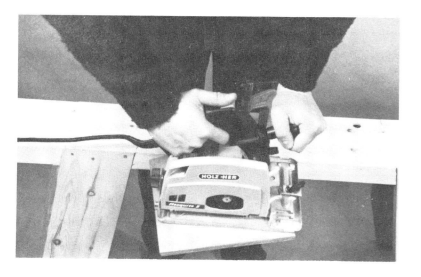

Cutting Arcs or Circles

Arcs or circles can be cut with a series of tangent cuts (Illus. 274). They can also be cut with a Thorsness™ (Illus. 275) or Laser™ blade. Take light cuts and work carefully when making circles or arcs. Guides or templates may also be helpful (Illus. 394–396).

Illus. 274. This arc was made with a series of straight cuts tangent to the arc.

Illus. 275. This Thorsness blade can cut an arc or circle in solid stock. Other blades such as the Laser can be used for particle board and other sheet stock.

Cutting Joinery

The portable circular saw can be used to cut joints such as the dado, lap, mitre, and rabbet joints. These joints require careful setup of the saw and exact layout of the joints. If you are familiar with the saw, the joint quality will be improved. Cutting joinery or other work that demands accuracy with an unfamiliar saw affects hand–eye coordination. Practice on scraps before you attempt the job. This will improve quality considerably.

For best results, cut joinery with a finer blade than you would use for general work. This will minimize tear-out. With some joinery, the work must be completed with a chisel or other hand tools.

Rabbet Joint The rabbet is an L-shaped channel along the edge of the work. On thick stock, make two cuts, one on the face of the work and one on the edge. On thinner stock, make the rabbet with a series of cuts on the face of the work because it is difficult to balance the saw base on a thin edge.

Begin either type of rabbet cut with a careful layout of the workpiece. Set the saw-blade depth of cut from the face of the work. It may be helpful to use a straightedge or guide to control the path of the saw (Illus. 276–278). For rabbets cut along the grain, the rip fence can be used to guide the saw. Test the setup on a scrap and make any necessary adjustments. Make sure that the work is clamped securely. Keep clamps away from the saw's path.

On thick stock, make a single cut on the face of the work along the layout line (Illus. 279–284). Lift the guard before starting the cut. Do not force the saw. If the blade stalls, it could veer off course.

Make the second cut on thick stock from the edge of the work. Set the blade depth from the edge of the work. The arc of the saw blade should be even with the bottom of the kerf. Set the fence to control the path of the blade. Lift the guard, turn on the saw, and begin cutting when the saw reaches full speed. Hold the saw securely to keep the blade on course. Push

Illus. 276. With the saw positioned in the guide, adjust the saw blade depth to the desired rabbet.

Illus. 277 (above left). A series of cuts with the saw forms a rabbet on the end of the workpiece. Illus. 278 (above right). With careful layout, you can cut an excellent-quality rabbet joint with a portable circular saw.

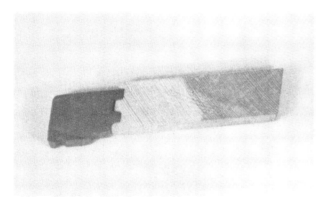 

Illus. 279 (above left). This replacement sill has two parts. Sometimes the darker piece must be rabbeted to accommodate a stone sill. Illus. 280 (above right). This is the typical rabbet that must be cut on the front sill piece.

Illus. 281. Make the first cut through the face of the stock. Tilt the blade so that it is parallel with the edge of the sill.

Illus. 282 (above left). Work carefully as you complete the cut. It is possible for the opening in the shoe to fall down on the work. Illus. 283 (above right). Make a second rabbet from the edge of the workpiece. Again, clamp the stock and tilt the blade.

Illus. 284. The completed rabbet cut. Good results were achieved through careful sawing, stock clamping, and accurate saw setup.

the blade clear of the work and shut off the saw. Lift the saw off the work after the blade comes to a complete stop. Inspect the rabbet. Some cleanup work with a chisel may be necessary.

On thinner stock, make a series of cuts to remove the waste stock. Use a chisel to trim out any excess wood.

Dado Joint A dado joint is a U-shaped channel with or across the grain. Dadoes made with the grain are sometimes called groove joints.

To make a dado, lay out the joint on the work and adjust the blade exposure to the dado depth. Use a straightedge to guide the saw. For some grooves, the fence can be used to guide it. Make the shoulder (face cut) the same way you would make a face cut in a rabbet (Illus. 285). Move the straightedge and make a second shoulder cut on the other side of the dado (Illus. 286).

When this cut is complete, remove the waste wood between the shoulder cuts with a series of cuts. You can use a straightedge if you want to, but it is not

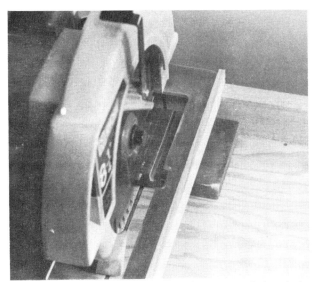

Illus. 285. Cut the sides or shoulder cuts of the dado first. Use a guide to control the saw.

Illus. 286. Both shoulders have been cut. Now make a series of cuts to remove any remaining stock.

Illus. 288. Begin the lap joint with a careful layout. The X indicates what stock should be removed.

necessary. As long as the saw blade does not contact either shoulder, the cuts do not have to be perfectly straight. Check the fit with the mating part (Illus. 287). Make any needed modifications. Clean the dado up with a router plane or chisel, if needed.

bottom of the layout line, and lock the saw securely. You can use a guide to control the shoulder cut (Illus. 289).

Illus. 287. A tight-fitting dado joint is obtained through careful layout.

Illus. 289. Use a guide to make the shoulder cut through the face of the work.

Lap Joint A lap joint is either a corner or intermediate wood joint. A corner or end lap has two identical parts. The cut looks like an elongated rabbet. Intermediate laps have one part that looks like a rabbet and one that looks like a dado. Cross laps have two identical parts that look like dado cuts.

To make an end-lap joint, lay out the mating parts and mark the sides on which the stock is to be removed (Illus. 288). Set the blade depth to cut the

Now, move out to the end of the workpiece and make a series of cuts across the work (Illus. 290). As you complete each cut, move back towards the initial shoulder cut. Work carefully. The saw blade could attach itself to small scraps and kick back. Clamp stock securely. Clean up the cut with a chisel if necessary (Illus. 291).

Cut the mating part in the same way or use a Laser™ blade. Make the shoulder cut the same way (Illus. 292). Then move the saw with a Laser blade side to side across the cut (Illus. 293). This will remove all stock smoothly (Illus. 294). Test the fit between the parts and make any needed adjustments (Illus. 295).

Illus. 290. You can remove the remaining stock with a series of cuts. Start at the end and work towards the shoulder.

Illus. 291. Use a chisel to clean up the lap joint and smooth the surface.

Illus. 292. Also start the mating part with a shoulder cut.

Illus. 293. Use a Laser blade to remove the remaining stock. The blade travels back and forth across the work, removing stock.

Illus. 294. The Laser blade leaves the joint perfectly smooth with no need for cleanup.

Illus. 295. Check the fit between the two parts to see if additional cutting is required.

An intermediate lap joint has one piece that looks like an end lap. It is cut the same way as an end lap. The mating part looks like a dado. Use the end lap to lay out the intermediate lap joint (Illus. 296). The intermediate lap joint requires two shoulder cuts (Illus. 297) and a series of cuts between the shoulders (Illus. 298). The depth of cut is the same for both parts (Illus. 299 and 300). This cut is made much like the dado cut discussed previously.

The cross-lap joint has two identical cuts similar to the dado. These are made in the same way as the dado. Always check your setup on a piece of scrap to

Illus. 296. An end lap part can be used to lay out an intermediate lap joint.

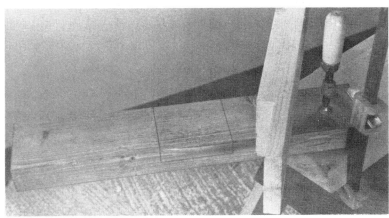

Illus. 297. Make the two shoulder cuts first. Use a guide to control the saw for greater accuracy.

Illus. 298. Make a series of cuts between the shoulders to form the intermediate lap.

Illus. 299. Clean up the intermediate lap cut with a chisel. Smooth the stock carefully.

Illus. 300. Test the fit between the mating parts. Make any needed adjustments.

make sure that you have a tight fit. Remember to mark the side of the work on which the stock is to be removed.

A dovetail lap joint is an intermediate lap joint with a dovetail configuration. The end-lap part is laid out like a typical dovetail. Cut the shoulders first (Illus. 301), and then make the angular face cut (Illus. 302). A hand saw must be used to complete the cuts where they meet (Illus. 303 and 304). Now cut away the back side of the piece like a typical end-lap joint (Illus. 305). After the wood is cut away (Illus. 306), trim the piece with a chisel (Illus. 307 and 308).

Use the end-lap part to lay out the mating joint (Illus. 309). Make your layout lines carefully and exact. You can use a straightedge to guide both shoulder cuts. Remove the remaining stock with a series of cuts, or use a Laser blade (Illus. 310 and 311). Test the fit between the parts (Illus. 312), and make any needed modifications with a chisel.

Mitre Joints Mitre joints are inclined end, edge, or face cuts. They are most commonly used to hide end grain on window frames and door trim. These cuts give the door or window a wrapped look. A complete discussion of mitres can be found in Chapter 5.

When you make an end mitre, the blade has to be tilted to the mitre angle. This angle is usually 45 degrees, but other angles can be used. The cut is a crosscut with the blade inclined. A straightedge guide can be used to control the cut, if desired. Some woodworkers simply use a square to guide the saw.

Make sure that the work is held securely while you are making the cut. If the cut is in the center of the work, make sure that both ends of the piece are supported. This will prevent the blade from binding. Check the cut with a sliding T bevel or combination square to determine how accurate it is. For best results, make a test cut in a piece of scrap.

Cut face mitres with the blade set perpendicular to the saw base. Use a mitre guide or straightedge to control the saw. The mitre is usually set at a 45-degree angle across the face of the work, and the cut is made the same way as a crosscut. Keep the saw against the guide for best results. Test the cut with a sliding T bevel or combination square.

When you are making up a frame using parts with end or face mitres, the angle of the cut is important to a good fit. Also, all mating parts should be equal in length. If one part is slightly longer, the mitres will not mate. They will be open on the inside of the frame. A quality blade is also essential. A blade with little tension may flutter while the cut is being made. This will cause saw marks or ridges on the mitre cut.

Edge mitres are actually bevel rip cuts. Set the blade at the desired angle (usually 45 degrees) and lay the cut out on the work. For some edge mitres, a straightedge is used to guide the saw. For others, the rip guide can be used.

Check the edge mitre the same way you would an end mitre. Remember to follow safe rip procedures when cutting it. A saw with a splitter may be useful for this cut.

Compound Mitre A compound mitre is a mitre with two angular cuts. When stock is tilted up at an angle, such as in a hopper cut, two angles are cut. These angles are derived from a mathematical formula. A chart is provided in Chapter 9 (page 239). Use the numbers in this chart to set the angle of your saw blade and the angle of the cut.

Practice on scrap when cutting compound mitres. Test the fit to be sure that the angles are correct. The compound mitre is frequently used in trim carpentry. Many types of cove moulding require cuts with a compound mitre.

Illus. 301. Cut the shoulders on the dovetail first. These cuts are made through the edge of the work.

Illus. 302. Make the angular face cuts next. When the face and shoulder cuts meet, stop the saw.

Illus. 303. Complete the cut with a hand saw.

Illus. 304. Now make the shoulder cut using the layout line.

Illus. 305. Use a guide to make the shoulder cut more accurate. Clamping also helps improve accuracy.

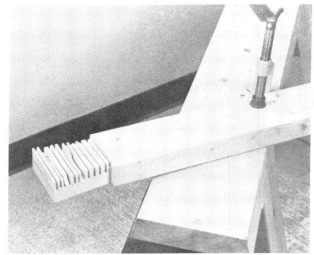

Illus. 306. The series of cuts made with the saw make chiselling easier.

Illus. 307. Work from the end towards the shoulder when chiselling. Work carefully to get best results.

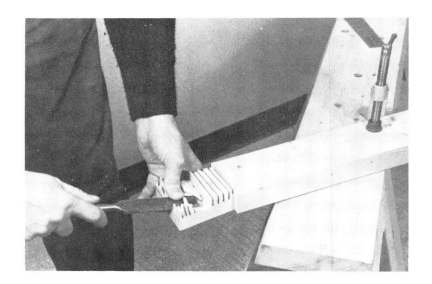

Illus. 308. Careful chiselling will produce a smooth lap surface on the dovetail.

Illus. 309. Use the end lap portion to lay out the mating intermediate portion. Make the layout carefully for best results.

Illus. 310. Make the shoulder cuts first and then use the Laser blade to remove the remaining stock.

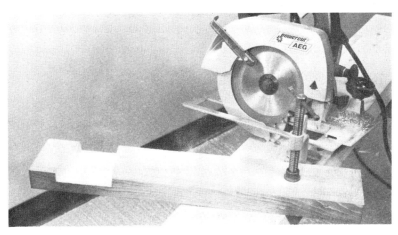

Illus. 311. The mating part has been completed with the Laser blade. Note the smooth bottom on the cutaway portion.

Illus. 312. Use a chisel to adjust the parts for a perfect fit. These pieces were pushed together with hand pressure after being trimmed.

Cutting Notches in Framing Members

Framing members are notched to either clear plumbing or electrical supply lines or to accommodate mating parts. Rafters and stair horses are notched for the latter reason.

The notches in a rafter rest on the plate. These notches are commonly known as bird's-mouths. The opposite end has a face mitre cut on it which allows it to rest snugly against the ridge board or ridgepole.

The notches in a stair horse are cut to accommodate the tread and riser. Mitre cuts made at both ends of the stair horse allow the stair horse to rest firmly on the floor and against the floor joist or header.

You can cut notches for pipes and wires in three different ways. Narrow notches are usually made with two face cuts slightly beyond the layout line (Illus. 313). This makes the back of the cut even with the layout line. The piece is then broken out with a rip hammer or a chisel (Illus. 314 and 315). Remember, the arc of the blade does not allow a straight cut.

Some woodworkers prefer to go in on the edge of the work with the blade set to the desired depth (Illus. 316 and 317). This piece can also be broken out with a chisel or hammer (Illus. 318 and 319). The difference between the cuts is that the edge cuts do not go beyond the layout line. A piece cut this way will be stronger. Large notches can weaken framing members. Some carpenters nail a reinforcing member along the cutout area to strengthen it.

Some carpenters make this narrow notch by extending the blade to full exposure and coming in at an angle (Illus. 320). This is a compromise between the two methods just described. The notch is cut to the layout line. The bottom cut will be relatively straight (Illus. 321) because the arc of the blade contacts the work at an angle (Illus. 322). The drawback to this method is that the entire blade is exposed for this cut,

Illus. 313. Notches in framing members can be made with two face cuts slightly beyond the layout line.

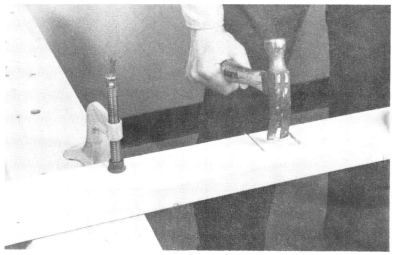

Illus. 314. Then use a rip hammer to break out the stock that's between the saw cuts.

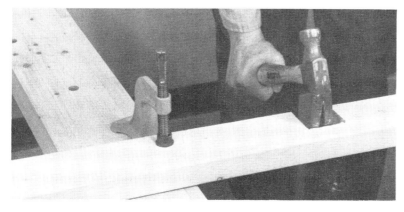

Illus. 315. Do the rough trimming with the rip hammer. You can use a chisel to make a smoother notch.

Illus. 316. Notches can also be cut in from the edge of the framing member. Set blade depth to the depth of the notch.

Illus. 317. Both cuts have a flat bottom. This is because the blade went through the cut instead of stopping near the layout line, as in Illus. 313.

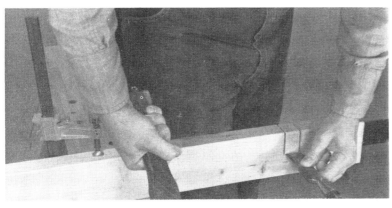

Illus. 318. A chisel can be used to break out the stock between the notch cuts. For coarser work, you can also use a rip hammer.

Illus. 319. A smooth bottom results when a chisel is used to form the notch.

Illus. 320. In some cases, the notch cut is made at an incline with the blade depth at its maximum setting.

Illus. 321. The bottom of the cut is relatively straight because of the way the saw had been inclined during the cut.

Illus. 322. Break out the notch with a rip hammer.

so there is a greater chance of kickback. This technique should only be used by experienced woodworkers and with great caution.

Wide notches may require a pocket cut along the bottom. This is because it would take too long to break out the piece with a chisel, and the notched area may not be uniform. Cut the ends of a wide notch in one of the ways just described. Then make the pocket cut between the cuts (Illus. 323). Saw the corners free with a hand saw (Illus. 324 and 325). In some cases, the piece is broken out with a hammer and the corners are squared-out with a chisel.

Rafters are face-mitred at one end. You can use a guide for this cut or make it freehand (Illus. 326). Lay this cut out carefully. It must butt snugly against the ridge. If it does not fit securely against the ridge, the strength of the roof system will be diminished greatly.

As already mentioned, the bird's mouth is a notch

Illus. 323. When a long notch is desired, make a pocket cut between two shoulder cuts.

Illus. 324. Use a hand saw to finish cutting the corners of the notch.

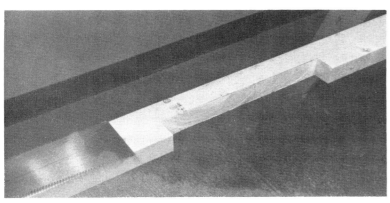

Illus. 325. Cutting the notch in this manner is much faster than chiselling away the stock between the shoulder cuts.

Illus. 326. Cut the face mitre on the end of this rafter with a guide. A guide can greatly improve the quality of a cut.

near the opposite end of the rafter that is made with two cuts. Lay out the position of the bird's mouth accurately. Make the cuts from the same edge of the work (Illus. 327–329). Complete the cuts with a hand saw or reciprocating saw; they should not go beyond the layout lines. If they do, the strength of the framing member will be reduced.

Stair horses are face-mitred at both ends. The angle of the cuts depends on the incline of the stairway (Illus. 330).

Notches are cut to hold the treads and risers (Illus. 331–342). Again, these cuts should not go beyond the layout line. Complete them with a hand saw or reciprocating saw.

When cutting notches, remember that the larger the diameter of the blade and the greater the blade exposure, the smaller the arc at the end of the cut (Illus. 343 and 344). This means that a 10-inch blade with full exposure will cut a notch for a bird's mouth without weakening the rafter as much as a 7¼-inch blade.

Many carpenters use a larger diameter blade when cutting notches. This minimizes or eliminates hand-cutting at the junction of the two cuts. However, the cutting procedure is not as safe because of the increased blade exposure.

Trimming Doors

Whenever carpeting is installed for the first time, most doors rub on the carpet. Trimming door bottoms is a common job done with the portable circular saw.

Planning is essential when trimming doors. First, determine how much stock must be removed. You can make a cutting jig to control the saw or use a straight-edge.

Be sure to mark the bottom of the doors when you remove the hinge pins. It is not uncommon for a car-

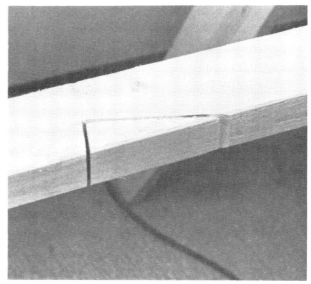

Illus. 327 (above left). Make the two cuts that form the bird's mouth from the same edge of the rafter. Illus. 328 (above right). Some stock remains in the corner where the two cuts meet.

Illus. 329. Cut the stock remaining in the corner with a hand saw or sabre saw.

Illus. 330. Mitre cuts are made at both ends of a stair horse. Note also the bevel rip on the framing member beneath the stair horse.

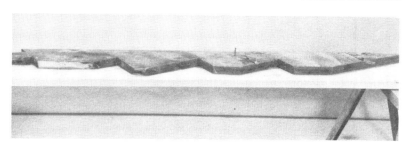

Illus. 331. In some cases, you can use the old stair horse or stringer to lay out the new one.

Illus. 332. Note that the original cutouts are not square; this makes it easier for the treads to hold water.

Illus. 333. Lay out the treads and risers carefully. You can fasten stops to the square to increase the accuracy of your layout.

Illus. 334. A small amount of stock remains in the corner of the stair cuts. This is due to the arc of the blade.

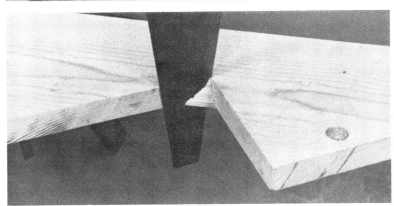

Illus. 335. Saw away the remaining stock with a hand saw.

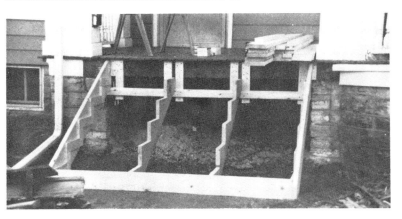

Illus. 336. Extra reinforcement will make these stairs very rigid even when a refrigerator or piano is carried down the steps.

Illus. 337. The outer horses, which have cleats glued and nailed to them, are more rigid because no material has been cut away.

Illus. 338 (above left). Making well-made stairs like these will present you with many challenges when you use your portable circular saw. Illus. 339 (above right). The lower braces require compound cuts. Cuts like these require careful measurement and layout.

Illus. 340. Make an end mitre cut at both ends of the railings.

Illus. 341 (above left). Use a sliding T bevel to set the saw to the correct mitre angle. Illus. 342 (above right). Adjust your saw carefully and test the setup on a piece of scrap before cutting the railing.

Illus. 343. This 10-inch saw leaves a small scrap in the corner of this notch.

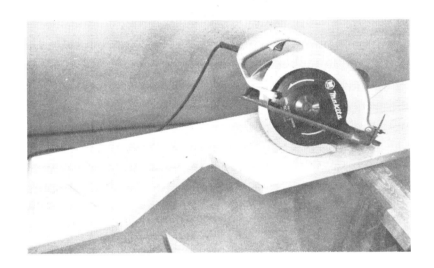

Illus. 344. This 16-inch saw leaves an even smaller scrap in the corner of the notch.

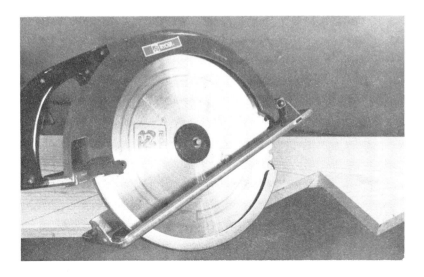

penter to trim the wrong end of the door if the door is not marked, or if it is marked incorrectly. Set the door on a pair of sawhorses or another firm work surface. Clamp it securely. Use clamp pads to protect it. Mark the cutting line carefully (Illus. 345). It is a good idea to score the face veneer with a utility knife (Illus. 346); this will control tear-out problems.

Many carpenters also tape over the scored layout line (Illus. 347). This helps hold the wood fibres down during the cut (Illus. 348–350). Remove the masking tape carefully after completing the cut (Illus. 351). If the tape is anchored securely, it can actually lift the wood fibres if it is pulled off carelessly (Illus. 352).

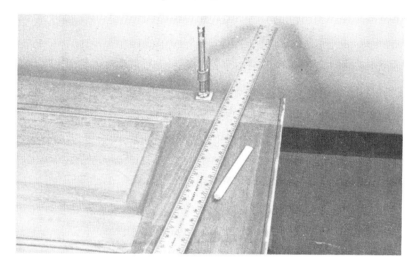

Illus. 345. Lay out the amount of stock to be removed carefully. Make sure that you are marking the bottom of the door.

Illus. 346 (above left). Score the face veneers and short grain with a utility knife. This will minimize tear-out. Illus. 347 (above right). Tape can also be useful in controlling grain tear-out. You actually saw through the tape.

Illus. 348. Clamp a guide along the cutting line. Use clamp pads to protect the door.

Illus. 349. The saw base rides along the guide. Keep the shoe firmly on the base.

Illus. 350. After you have completed the cut, lightly plane and seal the bottom of the door.

Illus. 351. Lift the tape carefully to avoid pulling up the wood fibres.

Illus. 352. Tape can lift up the wood fibres if it is pulled off carelessly.

Cutting Nonwooden Materials

Metal, stone, glass, ceramic and masonry can be cut with a portable circular saw. Generally, iron and steel are cut with an abrasive-type blade similar to a grinding wheel. This abrasive blade is usually made of aluminum oxide. Nonferrous metals are sometimes cut with a blade with special teeth. Stone, glass, ceramic, and masonry are usually cut with an abrasive wheel made of silicon carbide. Specialty diamond-grit abrasives can also be used. These cutters require water as a coolant. If you use an electric saw with coolant, there is a chance of electric shock. Wear rubber boots and rubber gloves and protect the circuit with a ground-fault circuit interrupter.

When you are cutting materials other than wood, most saws need special washers for the abrasive or diamond blade you will be using. In addition, some worm-drive saws need a specific adapter so that a round arbor-hole center can be used. This eliminates the positive drive and minimizes the chance of the abrasive blade shattering. Consult the owner's manual or manufacturer's catalogue before cutting nonwood materials. This will enable you to use the correct accessories and work as safely as possible. More information about blades can be found in Chapter 3.

Safety Guidelines Nonwood materials include glass, metal, ceramics, stone, and concrete. Each of these materials requires a special blade or cutter. Observe any warnings or precautions suggested by the manufacturer, and only use a blade or cutter for its intended purpose.

Grinding-type cutters are usually bonded with a fiberglass mesh that minimizes the chances of their shattering. Always guard at least one-half or more of the cutter. Never use a grinding-type cutter that has been dropped.

When you install a new cutter, let it run for one minute or longer before you begin cutting. Listen for any noise that would indicate a problem. Also make sure that the speed of the saw (rpm's) is compatible with the cutter.

Stand to the side of the cutter; this will keep you out of the path if it were to shatter. Avoid twisting the cutter in the kerf. Side stress is very hard on it. Do not force the cutter into the material; let it do the cutting. Too much force can damage the cutter or cause premature wear.

When cutting concrete, slate, or other masonry products, remember that each cut or pass should be no deeper than ⅜ inch. When cutting metal, expose only enough cutter to get through the material. Also, be aware of fire hazards. Some cutters generate sparks and hot chips which could easily ignite wood dust or chips.

Always protect yourself when cutting nonwood materials. Use protective glasses and a face shield. Protect your hearing with ear plugs or muffs. Remember to protect your lungs. The dust generated by masonry and metal products can be very harmful. Minimize your risk whenever possible.

Cutting Metal Steel and some nonferrous materials are cut with an aluminum oxide cutter or blade. Mount the appropriate blade according to the manufacturer's instructions, using the appropriate accessories. Set the depth of cut according to the specifications of the blade manufacturer. It may vary from one blade manufacturer to another.

Clamp the stock securely and wear the appropriate protective equipment when cutting metal (Illus. 353–359). In addition to safety glasses, this equipment may include gloves, hearing protectors, and a dust mask. When in doubt, use all of this protective equipment.

Turn on the saw and let it come up to full speed. Guide the abrasive blade into the work. Push only as fast as the abrasive blade will cut the work. Forcing the blade will cause excessive wear, and may cause it to shatter. Abrasive blades will cause sparks to fly in the same way as a grinding wheel. Keep combustible material and debris clear of the cutting area. It is a good idea to keep a fire extinguisher nearby when using an abrasive-cutting blade.

Some metal-cutting saw blades have teeth similar to those on a wood-cutting blade (Illus. 360). These blades have a tooth size and shape designed for a special job. Make sure that the blade you install is

Illus. 353. Clamp the stock securely when sawing with an aluminum-oxide cutting wheel.

127

Illus. 354. Sawing with a cutting wheel is the same as sawing with a blade. Keep cutting-wheel exposure to a minimum.

Illus. 355 (above left). Some flash of metal will cause the cutoff to cling to the work. It can be broken off easily by hand.
Illus. 356 (above right). Cut channel iron across the entire cross section when possible.

Illus. 357 (above left). Break away the cut off with a twisting motion. Illus. 358 (above right). Notches can be made with two or more cuts. Make the first cut from the end.

Illus. 359. The second cut from the edge completes the notch.

Illus. 360. The saw blade was designed for cutting aluminum.

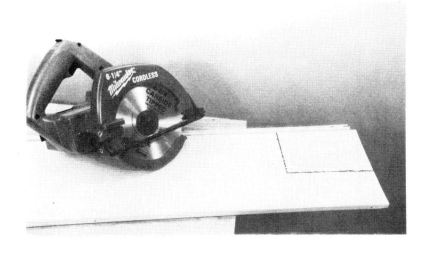

Illus. 361. Lay out the stock the same way you would lay out the wooden parts.

Illus. 362. Clamp the stock to a sawhorse or other work support before sawing.

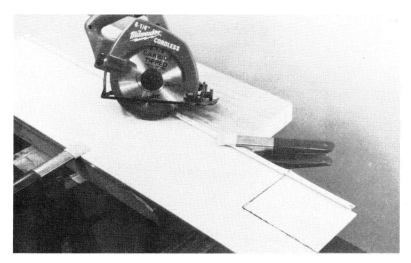

Illus. 363. Cut the metal the same way you cut wood. Observe all safety precautions.

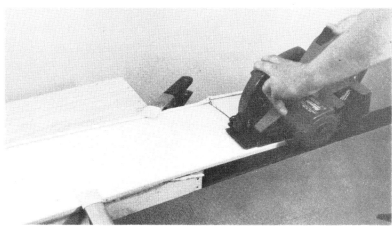

correct for the job. In some cases, these blades require a particular type of lubricant. Make sure that this material is available.

Clamp the metal securely for sawing. Set the depth of cut according to the manufacturer's instructions. Saw the metal the same way you would saw wood (Illus. 361–363). Wear the appropriate protective equipment and observe the same safety rules you would for cutting wood.

Cutting Stone and Masonry Stone and masonry are usually cut with a silicon-carbide blade similar to a grinding wheel. Make sure that the blade is appropriate for the material being cut. Set the blade exposure according to the manufacturer's instructions, and wear the appropriate protective equipment, which must include a dust mask to protect your lungs from the harmful dust.

Before cutting the work, make sure that it is being held securely. If the stone is larger than the cut of the saw, cut from all sides and break it with a mason's hammer. Take light cuts when cutting masonry (Illus. 364–366). Avoid twisting the cutter in the kerf.

Some diamond wheels require the use of a coolant. The coolant contains the dust, speeds the cutting, keeps the wheel cool, and reduces the wear on the wheel. When water is used as a coolant with an electrical saw, be aware of the possible electrical shock hazards. Follow all manufacturer's recommendations when using diamond wheels (Illus. 367–369). Make sure that the saw is grounded or double-insulated and plugged into a ground-fault circuit interrupter.

Cutting Glass and Ceramics Glass and ceramics are cut with a diamond wheel and coolant (Illus. 370–372). Large glass- and ceramic-cutting jobs should be done with glazier's tools, but many times glass or ceramics must be cut to fit or accommodate an electrical outlet or other supply line. Work only as fast as the wheel will cut, and keep the coolant flowing. Cut arcs with a series of tangents. Do not twist the wheel in the kerf. Abnormal stress on the glass can

Illus. 364 (above left). Light cuts ¼ inch deep are the best ones to make when cutting masonry. Illus. 365 (above right). Hold the saw firmly. Do not rock or twist it.

Illus. 366. A second cut has been made on this concrete block. Additional cuts will separate the pieces.

Illus. 367 (above left). A diamond blade is being used to cut this slate. The slate is being cut in one pass. It is ⅜ inch thick. Illus. 368 (above right). Cut only as fast as the blade will cut. Wear a dust mask to protect your lungs.

Illus. 369. Some diamond blades can only be used with water. Be sure to check manufacturers' specifications before you begin cutting.

Illus. 370. A special coolant bottle is attached to this cordless saw used for cutting glass.

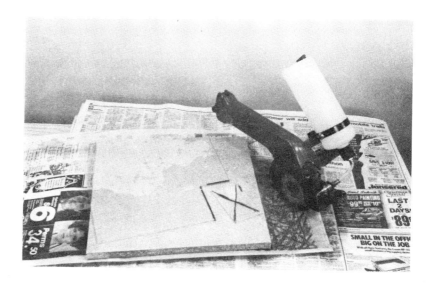

Illus. 371. A notch has been cut from the corner of this piece. Use plenty of coolant or the glass is likely to crack.

Illus. 372. Pocket cuts are also possible when you are cutting glass. Newspapers help blot up the coolant. You can also do the sawing over a tub, which the coolant will drip into.

cause it to shatter. Breaking glass or making incorrect cuts can be very expensive. Work carefully and measure accurately.

Stationary Saw Guides and Devices

In addition to the number of portable-cutting guides that attach to the work (presented in Chapter 4), there are a number of stationary guides and devices. The guides provide a track or engagement in or on which the saw travels to make a straight cut. The work butts or clamps against a stop for cutting. These guides and devices are discussed by their trade names.

The saw guides and devices presented here do not fit all portable circular saws. For best results, check the manufacturer's specifications before purchasing any of them.

***Miter Maker*™** The Miter Maker controls the cut of a portable circular saw and transforms it into a machine similar to a motorized mitre box (Illus. 373–375). Move the fences to the desired angle to control the angle of cut. Set the blade perpendicular to the base; it has no pivoting capability.

A length stop is provided with the tool with which the saw can cut several pieces to uniform length. This device makes mitring easier with a portable circular saw, but some time is required to set the fences in the correct position (Illus. 376–384). As with any machine, you should check your setup on scrap before actually cutting the work.

***Saw Shop System*™** The Saw Shop System controls the saw with an edge guide. The edge guide also acts as a clamping mechanism. The Saw Shop System is mounted on a piece of sheet stock (Illus. 385–387).

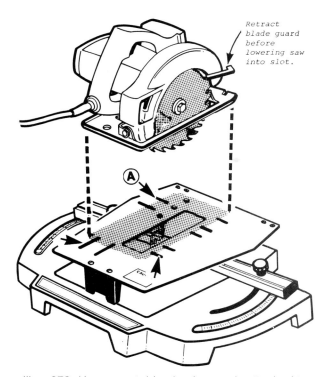

Illus. 373. Here a portable circular saw is attached to a Miter Maker, which is a portable circular saw accessory that enables the saw to perform the same functions as a motorized mitre box.

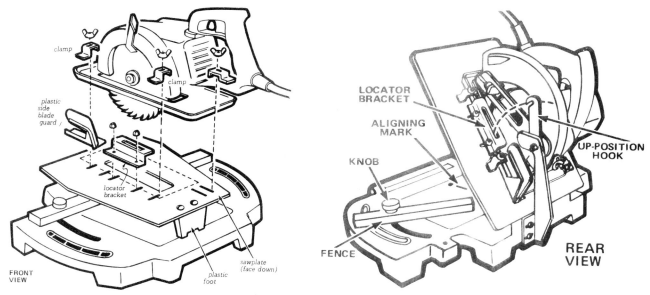

Illus. 374 (above left). The Miter Maker will also accommodate a worm-drive saw. Illus. 375 (above right). The Miter Maker has a hook that holds the saw in the up position. To adjust the Miter Maker, hook the saw in this position.

Illus. 376. Clamp a work support to a pair of sawhorses. This will provide a work surface for the Miter Maker.

Illus. 377. Clamp the Miter Maker to the work support. This makes sawing safer and more accurate.

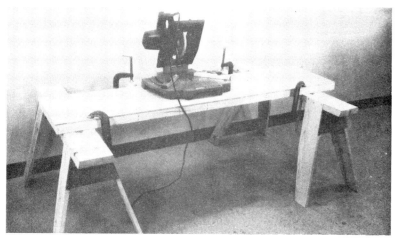

Illus. 378. To cut a mitre, move the fence to the desired angle. Tighten the knob on the fence to lock the setting.

Illus. 379. Turn the saw on and lower it into the workpiece. Your hands should never be under the plate on which the saw is mounted.

Illus. 380 (above left). After you have made the cut, turn off the saw, lift it, and lock it in the up position. Illus. 381 (above right). Cut the mating mitre against the opposite fence. Always lock the fence securely in position after making any adjustment.

Illus. 382. To make crosscuts, square and align the fences.

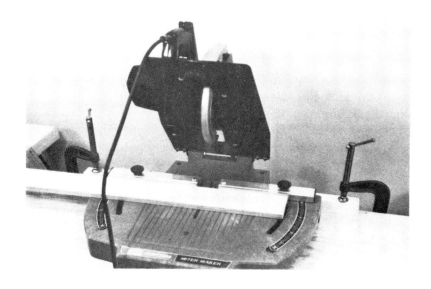

Illus. 383. Make the crosscut the same way you would make a mitre cut.

Illus. 384. Accurate crosscuts and mitres can be made easily with the Miter Maker.

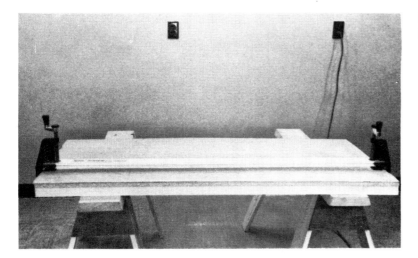

Illus. 385. Mount the Shop System on a work support or platform before using it.

Illus. 386. Clamp the stock under the guide bar by turning the cranks at each end. The layout line should be even with the clear plastic base.

Illus. 387. The saw rides along the guide bar and makes a straight cut.

The uprights hold the guide and act as a fulcrum point for clamping pressure (Illus. 388). Any layout line that lines up with the straightedge can be cut (Illus. 389 and 390). This jig is capable of cutting mitres and other joints (Illus. 391–393), as well as cutting circles (Illus. 394–396). When stock does not extend far enough to support both sides of the portable circular saw, tape a piece of scrap to the base of the saw. This keeps the saw perpendicular to the base.

Illus. 388. Line up a piece of stock and clamp it at an angle. This allows the Shop System to guide a taper cut.

Illus. 389. The saw rides along the guide bar and makes the taper cut.

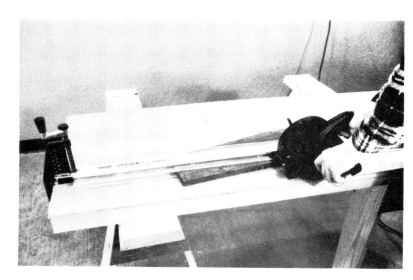

Illus. 390. Always keep the stock clamped with the Shop System. Blade depth should only be slightly greater than stock thickness. This minimizes the number of cuts made in the work support.

Illus. 391. Dadoes and other joints can be cut with the Shop System. Blade-depth adjustment is very important when cutting joints.

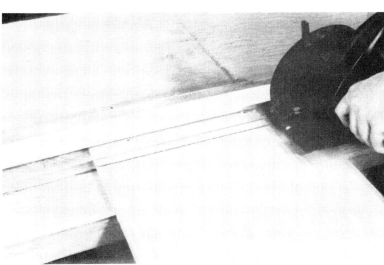

Illus. 392. Cut the shoulders first, and then remove the remaining stock by repositioning and clamping the work. Make a cut each time you move the stock.

Illus. 393. You will get a tight-fitting joint when you do the setup and layout carefully and accurately.

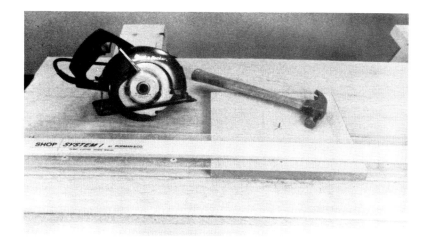

Illus. 394. To cut a circle, begin with a square blank. Line the edge up with the cutting guide and nail the plywood to the work support. Nail through the center of the plywood.

Illus. 395. Begin the circle by cutting off a corner. Then turn the work a few degrees and reclamp it. Make another saw cut.

Illus. 396. Through a series of tangent cuts, a circle is formed. Be sure to clamp the work for each cut.

Cutting and Finishing Jig With the cutting and finishing jig, you can use the portable circular saw like a radial-arm saw (Illus. 397 and 398) or Sawbuck™ to crosscut (Illus. 399 and 400) and to make joinery (Illus. 401 and 402). The jig pivots for angular crosscuts or mitres. In addition, you can tilt the blade on the saw to make compound-mitre cuts (Illus. 403 and 404). The tracks on the crosscutting jig are spring-loaded. After the work is positioned, they drop down onto the work.

Adjust the saw-blade depth accordingly (Illus. 405). Then turn the saw on and push it into the workpiece. When the cut is completed, the forward thrust of the saw lifts the tracks to their elevated position. This allows the telescoping guard to drop into position.

You can make angular cuts and compound-mitre or hopper cuts the same way. Make sure that the cut is laid out correctly before you begin. When in doubt, make a practice cut on scrap or the waste stock of the workpiece.

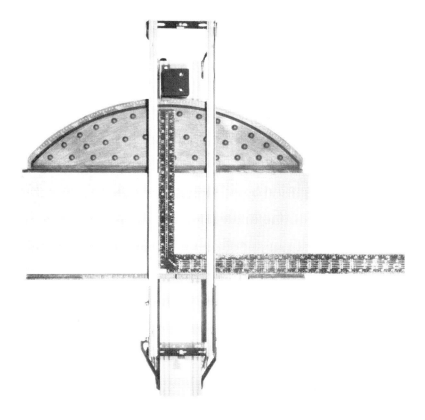

Illus. 397. The cutting and finishing jig has tracks and a plate on which a portable circular saw can travel. First square the tracks with the table.

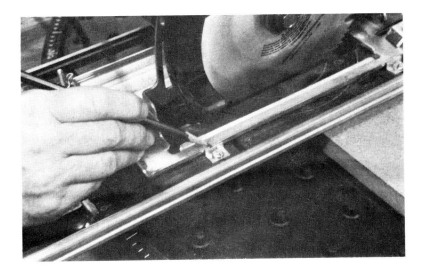

Illus. 398. Mount the saw on the plate and adjust it with the metal tabs. The plate holds the lower guard up.

Illus. 399. To begin a crosscut, position the saw. Turn it on and push it forward. The metal bars help hold stock on the table.

Illus. 400. As you complete the cut, the bars lift up and hold the saw well above and clear of the work area.

Illus. 401. You can cut lap joints using the cutting and finishing jig. Set the blade depth on scrap for best results.

Illus. 402. You can cut mitre joints by turning the jig for an angular cut. A scale on the jig makes angular setups easy.

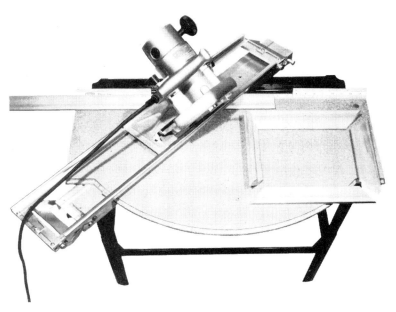

Illus. 403. With the blade tilted and the arm turned, compound mitres can be cut. Use the chart in Chapter 9 (page 239) to adjust the angles.

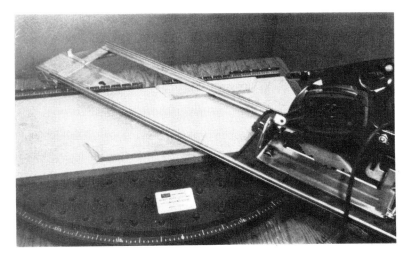

Illus. 404. On smaller stock, you can push scrap under the track to support it during the cut.

Illus. 405. The plastic block at the end of the saw's path helps you adjust blade depth. The blade should just touch the block when the work is under the rails. Raise it a little if you plan to make dadoes or other joints.

Siding Jigs Siding jigs are usually custom-made by people in the siding trade (Illus. 406–408). They are used to cut wood, vinyl or aluminum siding to length (Illus. 409 and 410). Since siding pieces are butted end to end on long runs, an accurate crosscut is needed.

The siding jig is usually built on a 2-inch plank. The jig has two L-shaped tracks mounted above the plank. The tracks are high enough to allow the work to pass between them and the plank. On some jigs, the tracks are made of hardwood; others use angle iron. The tracks should be rigid enough to resist deflection.

It is important that the tracks be perpendicular to the stop on the edges of the plank. This ensures that the crosscut pieces will have a square end (Illus. 411). You may want to make a siding jig for your saw. Plans for a siding jig can be found in Chapter 11 (pages 311–314).

Workmate™ Cutting Jig This jig is designed to be used with the Workmate™ folding sawhorse (Illus. 412) system. It is a two-part sliding system. The upper part clamps to the saw (Illus. 413 and 414) and controls the saw's path. A fence below the lower portion of the jig can, of course, cut a straight line (Illus. 415), and can be turned for mitres and compound mitres (Illus. 416 and 417).

Slide the stock under the jig and position it relative to the blade. Hold the stock securely against the fence and feed the blade into it (Illus. 418 and 419). The guide controls the cut, and the scrap will fall free. Support long pieces on a roller stand or saw horse (Illus. 421).

This cutting jig can only be used for crosscuts, mitres, and compound mitres. It is not intended for ripping operations.

Table-Saw Device Table-saw devices are available for the portable circular saw. These devices clamp the portable circular saw to the underside of a device that looks like a table saw. This device is equipped with a fence and mitre gauge, but no motor. The portable circular saw becomes the motor (Illus. 422–430). On some table-saw devices, the telescoping guard on the saw is used as a guard. Other devices come with a guard and a splitter identical to those on a table saw.

Illus. 406. This siding jig is designed for right-angle crosscutting in the field. This device is commonly used by siding installers.

Illus. 407 (above left). The saw rides in wooden rails that have been rabbeted. Angle-iron rails are sometimes used instead of wood. Illus. 408 (above right). Position the saw for cutting and adjust the blade height.

Illus. 409 (above left). Position the wooden bevel siding against the back of the jig. Adjust the cutting line to the saw kerf. Illus. 410 (above right). Turn on the saw and push it into the workpiece. The tracks control the blade's path.

Illus. 411. Panel stock can be squared up with this siding jig. The siding jig gives you shop-like quality in the field.

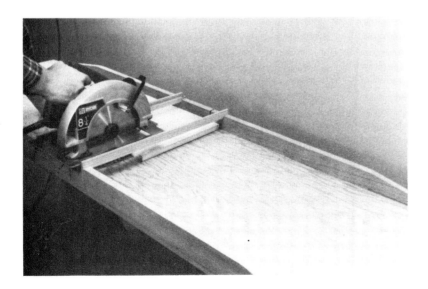

Illus. 412. This sawing jig can be used with a Workmate folding sawhorse. The jig extends above the bench just enough to allow the work to clear.

Illus. 413. The saw clamps to a piece that slides on the jig.

Illus. 414. Wing nuts are used to secure the saw to the sliding part of the jig.

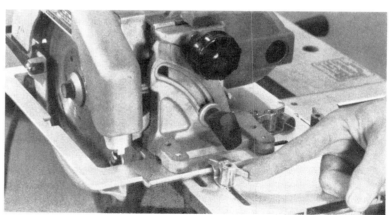

Illus. 415. You can adjust an angular scale for mitres. The opening in the jig shows the angle of the cut.

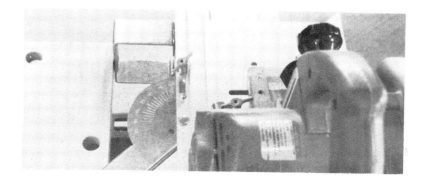

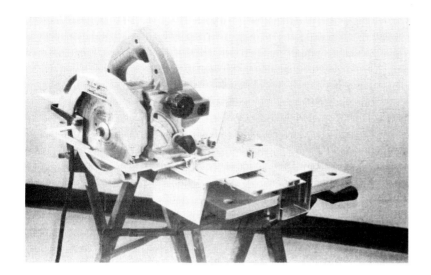

Illus. 416. The saw cuts a straight line due to the control provided by the jig.

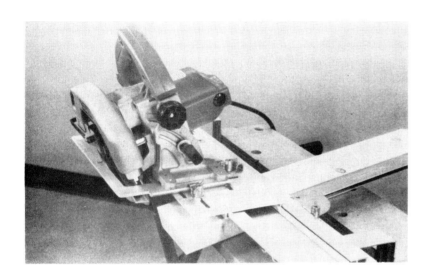

Illus. 417. You can make compound-mitre cuts easily with this jig.

Illus. 418. You can make crosscuts quickly and accurately with this jig. Stock is held against the fence while you are making the cut.

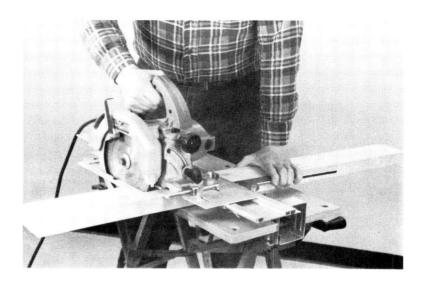

Illus. 419. Use conventional sawing practices when using the jig. Limit blade exposure to stock thickness.

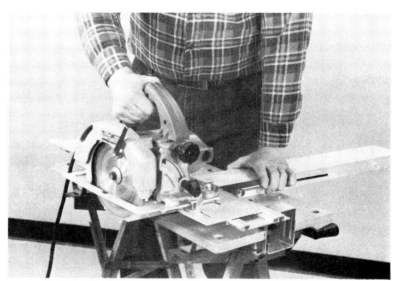

Illus. 420. Feed completely across the piece so that the lower guard returns.

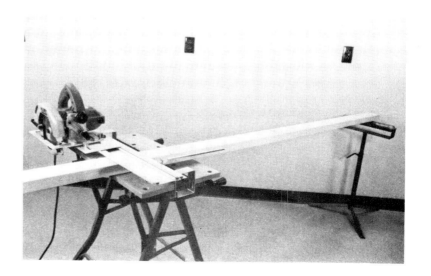

Illus. 421. Long pieces can be supported with a roller support or other device. Unsupported long stock could tip the Workmate over.

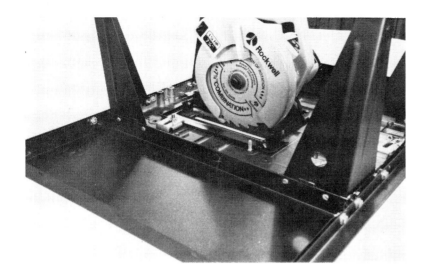

Illus. 422. First assemble the table-saw device and then bolt the portable circular saw to its underside.

Illus. 423. Special brackets and bolts hold the saw in position.

Illus. 424. Tighten the bolts securely after adjusting the blade.

Illus. 425. Measure the distance from the blade to the mitre slot at the front of the saw.

Illus. 426. Measure the distance from the same tooth to the mitre slot at the rear of the saw. When the distance is the same, the saw should be bolted securely in position.

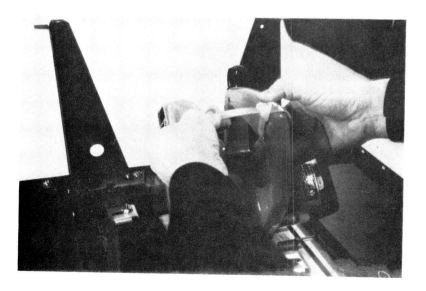

Illus. 427. Strap the switch on the portable circular saw to the on position.

Illus. 428. Plug the saw into the switch on the front of the table-saw device.

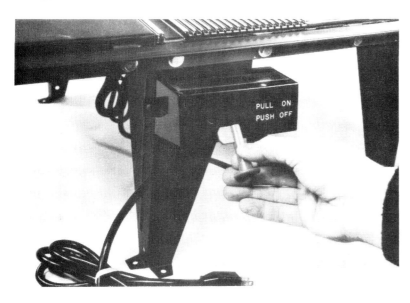

Illus. 429. A special key-type switch is used to operate the table-saw device. This keeps unauthorized people from operating the saw.

Illus. 430. Mount the guard to the table-saw device for protection. The lower guard on the portable circular saw is held below the saw table.

The device transforms a portable saw into a stationary table saw with which the stock can be cut while the saw remains stationary. This is useful for long rips and joinery cuts (Illus. 431–434). Observe safe table-saw cutting practices when using this device. They can be found in my book *Table Saw Techniques* (Sterling Publishing Co., Inc., Two Park Avenue, New York, New York 10016).

Some woodworkers make their own table-saw devices (Illus. 435 and 436). These devices perform the same functions but lack accessories such as the splitter and fence.

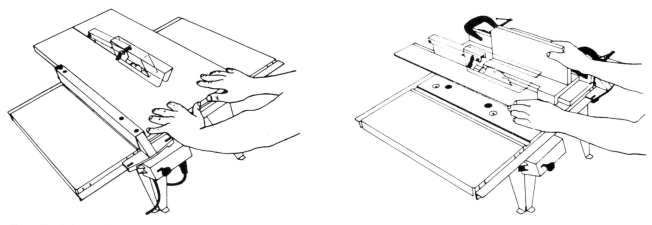

Illus. 431 (above left). Use the fence to control stock while it is being ripped. Clamp the table-saw device securely to a work support or bench. Illus. 432 (above right). A special push block can be used for narrow rips. This push block allows you to use the guard for this operation.

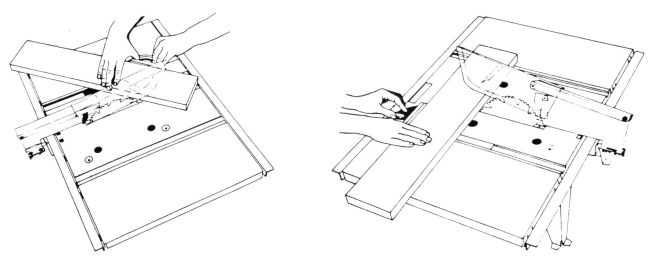

Illus. 433 (above left). The mitre gauge can be used for crosscuts or mitres. Hold stock firmly while making the cut. Illus. 434 (above right). You can also tilt the blade for end mitres and compound mitres. Use the mitre gauge to control stock for mitring and compound-mitring.

Illus. 435. This table-saw device was shop-made. It was photographed on the work site.

Illus. 436. On this device, the guard remains over the blade; a mitre slot is cut into the table.

Radial-Arm-Saw Device This device, commercially known as the Saw-Mite™ (Illus. 437), cradles the portable circular saw and allows it to be used like a radial-arm saw. The portable circular saw is capable of any common cut made by a radial-arm saw. The Saw-Mite can also be used with a router (Illus. 438). It is small enough for easy transportation and storage (Illus. 439).

Illus. 437. With the Saw-Mite you can use a portable circular saw like a radial arm saw.

Illus. 438. The Saw-Mite can also be used with a portable router. Consult the owner's manual for details.

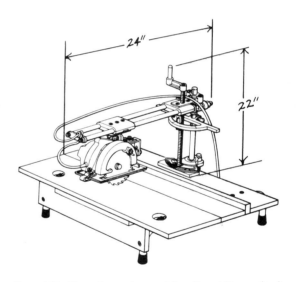

Illus. 439. The dimensions of the Saw-Mite make it portable and easy to handle.

This device and the table-saw devices make the portable circular saw work like a stationary saw. There are some problems when using them, however, especially when the portable circular saw is mounted upside down. In most cases, the portable circular saw used is not designed for this purpose. Also, sawdust drops down onto the motor. In addition, the saw may not be aligned with the mitre slot. This causes a heeling condition. A heeling condition on a stationary tool contributes to grain tear-out and kickbacks.

7
General Care and Maintenance of Your Saw

The portable circular saw is used under a variety of field conditions. These conditions can damage the saw if it is not maintained periodically. Dirt and grime are hard on the saw; they can damage the bearings, block the ventilation holes, and scratch the base. Clean the saw periodically. Brush or blow the dirt away from the ventilation holes and adjusting mechanism (Illus. 440). Wipe away dirt in lubricating cups and fill holes with a rag. Dirt sticks to oil and grease, so wipe up the lubricating material after lubrication.

Wax the base of the saw periodically to keep it running smoothly on the work. Inspect the base when you wax it. Remove any rust with steel wool or wet/dry abrasives. Remove any burrs on the base with a file or abrasives. Burrs on the base can scratch the surface of your work during a cut.

Inspect the base for warpage. Many bases become twisted or warped. This is usually caused when the saw is dropped. Thinner bases are more likely to twist. A twisted base makes it difficult to get an accurate cut, and is likely to cause the blade to bind.

Use two wood sticks to check for warpage. Place one stick at each end of the base. Sight along the base to see if the two sticks are in the same plane.

A warped base can sometimes be twisted back into a true plane with a parallel or C-clamp. Work slowly and carefully. You could twist it too far and be no better off. The best practice is to avoid dropping the saw. Keep it close to the ground where it cannot easily be knocked to the ground.

Inspect the electrical cord and the plug periodically. Repair any cuts in the protective outer insulation with electrical tape. Replace the cord any time you notice a cut that goes through the outer insulation. Virtually all portable circular saw manufacturers offer new cords that are ready for installation.

Also inspect the plug. If the grounding plug is missing, replace it. Under some conditions, the prongs on the plug will oxidize. In most cases the oxidation appears as a white or green powder on the prongs. This oxidation causes electrical resistance, which will cause a drop in voltage and amperage. Running the saw under this condition can cause the motor to burn out prematurely. Clean the oxidation off the prongs with a very fine (400–600-grit) wet/dry abrasive.

Inspect the mechanism that controls the base adjustment. Keep it clean. Remove any dirt or wood dust. Wax the mechanism to keep it moving freely.

Illus. 440. Keep dirt away from the saw with a brush or compressed air. When dirt accumulates, it can damage the bearings or electrical components of the saw.

Keep any threads on the clamping knobs or adjusting levers lubricated and free of corrosion. This will make it easier to adjust the saw.

Inspect the guard also. It should move freely in both directions (Illus. 441). Check it to make sure it retracts easily. The easier it retracts, the easier it is to get an accurate cut. Check it for ease of return. The guard should snap back. It should not have a lazy return. A lazy return can be caused by dirt in the working mechanism or a weak return spring. Identify the cause and repair it. A saw with a lazy return is unsafe and should not be used.

While inspecting the guard, move it from side to side to see if it will hit the blade. Look inside the guard for any signs of blade contact. Make any needed adjustments before using the saw.

Changing the Blade

Blade changes may be frequent or seldom, depending on how the saw is used. Saws used for general house-framing may use only one type of blade. This blade would be changed only when it is dull. Saws used for several purposes will require more frequent blade changes.

To change the blade, disconnect the saw from its power source (Illus. 442). Unplug the saw, or remove the battery if you are using a cordless saw.

Illus. 441. Check the guard movement periodically. It should snap back. A saw with a slow or lazy lower guard should not be used.

Illus. 442. Disconnect the power before changing the blade.

Some saws have a locking mechanism to hold the blade stationary while the arbor bolt is being removed. Push down on the button and slowly turn the blade until the lock engages (Illus. 443). Other saws are furnished with two wrenches (Illus. 444). Use one wrench to hold the arbor, and the other to loosen the arbor bolt. On some saws, you have to place the wrench that holds the arbor behind the blade. On other saws, you will engage both wrenches in front of the blade.

Next, remove the arbor bolt. On direct-drive saws, the arbor bolt usually has right-hand threads. On worm-drive saws, the arbor bolt has left-hand threads. Turn the arbor bolt slowly to be sure that you are turning it in the right direction; then remove it and the washer(s). Retract the guard and remove the blade (Illus. 445).

Illus. 445. Retract the guard to remove and replace the blade. Work carefully; even a dull blade can cut you.

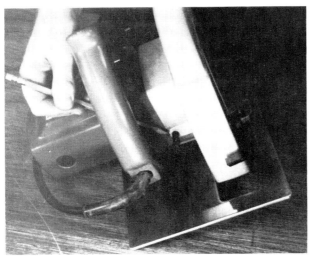

Illus. 443. Lock the blade with the lock mechanism if the saw has one.

Inspect the arbor washers for any imperfections. True them up on a stone if they have any burrs. The flat faces will make the blade turn in a truer plane.

Place the new blade over the arbor. Make sure the arbor hole is exactly the same size as the blade. If you are working with a worm-drive saw, the arbor hole will be diamond shaped. Make sure that it fits securely in

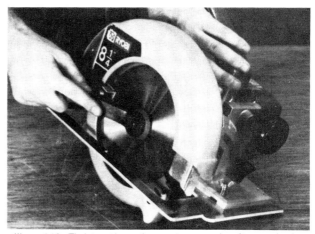

Illus. 446. Tighten the arbor nut securely after replacing the blade. If a spring-loaded washer is used, back up one-sixth of a turn after tightening. This will allow for the proper amount of slippage.

Illus. 444. On some saws you have to use two wrenches to remove the blade. Make sure that the wrenches are engaged properly before exerting any force.

place and that the teeth are pointing in the correct direction when you mount the blade.

Replace the arbor bolt and tighten it securely. Most direct-drive saws have a spring-loaded washer on the outside of the arbor washer (Illus. 446). This washer acts like a clutch and allows the arbor to slip if the blade pinches, which reduces the chance of a saw kickback. The spring does not have to rest perfectly flat against the blade.

The diamond arbor does not slip. This means that all power is transmitted to the blade. A kickback is more likely with this type of drive system. Tighten the arbor bolt securely on a diamond-drive saw. Also tighten the arbor bolt securely on a saw that has an electronic brake; the sudden stop of the blade can loosen the arbor bolt.

Adjusting the Saw

After you have installed a new blade, it is a good idea to adjust the saw for accurate cutting. One of the most important adjustments is to square the blade to the base. Move the base as close to the motor as possible and lock it in position. Use a square to check the angle (Illus. 447). Slide the square across the base until it touches the blade. Keep the square off the blade's teeth. Check the adjustment. There should be no light showing between the blade and square.

Illus. 448. Some saws have a stop that allows you to adjust the base perpendicular to the blade. This one also has a stop for a 45-degree angle.

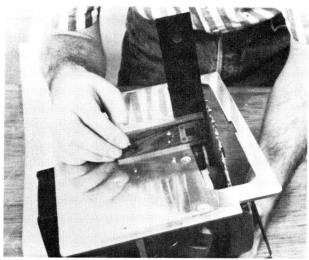

Illus. 447. Check the setting between the base and the blade with a square with the power disconnected.

Make the needed adjustment and lock the base in position. Check it again after locking the base to be sure of accurate adjustment. On some saws, there is a positive stop for the 90- and 45-degree settings. (Illus. 448). These settings can be adjusted with a pair of wrenches. The stops allow you to return to the 90- or 45-degree setting without checking it, which is a very useful feature because it allows for positive adjustment and saves time in the field. Adjust the 45-degree setting and stops using any square with a 45-degree angle. This is usually done after you have set the 90-degree stop.

On some saws, the base has an adjustable guideline. You can align the edge of the blade with this moveable part (Illus. 449 and 450). Place framing square against the blade to align it. The square should rest on the teeth that point towards the guideline. Tighten the adjusting screw to hold the guideline in position.

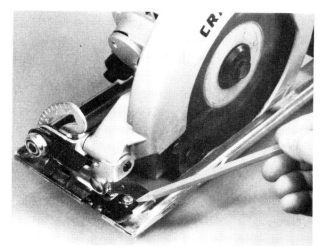

Illus. 449. The guideline or mark can be moved on this saw. This allows adjustment for various-size blade kerfs.

Illus. 450. This saw has the guideline cut out of the base. The insert goes in the opening around the blade. It helps reduce tear-out when you are cutting.

Test the guideline on a piece of scrap. Draw a straight line on the scrap and cut along the line using the guideline to control the saw's path. When the saw is about 10 inches into the scrap, shut it off. Allow the blade to come to a complete stop. Disconnect the saw and back up in the kerf. Check the relationship of the guideline to the saw kerf. Make adjustments if necessary. Some woodworkers prefer to set the saw guideline with the above method rather than using a square.

The guidelines on many saws are cut or filed into the base. These guidelines may not coincide with the saw kerf cut by the new blade, but they serve as an excellent reference point during the cut.

Some saws have more than one guideline. The extra guidelines may be for the other side of the kerf or for mitre settings (45-degree blade tilt). Your owner's manual can provide details.

On portable circular saws equipped with a splitter, it is important to check the relationship between the blade and the splitter. If the splitter is out of alignment, rip cuts will veer off the desired cut. Use a square to check alignment. Check both sides of the blade. Put the square against two teeth that point towards the square. The splitter should have equal clearance on both sides. When you check splitter alignment, make sure that the power is disconnected.

Lubrication of the Saw

Most portable circular saws require some form of lubrication periodically. For best results, follow the recommendations in your owner's manual. The following lubrication recommendations are based on a review of many owner's manuals. They may not be exactly correct for your saw. Remember to avoid overlubricating. Extra lubrication attracts dirt and dust and may discolor your work.

The arbor bearings on many saws require periodic lubrication. Other saws are lubricated once for life. A common method of lubricating the arbor is with a grease cup (Illus. 451), which is usually located on the back of the saw. It has a pipe that extends into the arbor. The cup is similar to the lid on a bottle except that it is deeper and has finer threads. This cup is filled with a special cup grease (Illus. 452) and

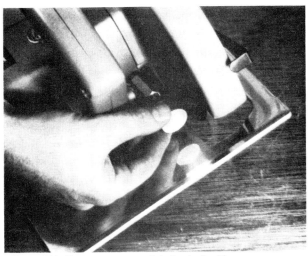

Illus. 451. Some saws are lubricated with a grease cup. Tightening the cup increases lubrication to the saw.

Illus. 452. When the grease cup is empty, remove it and fill it. Consult your owner's manual for the correct lubricating material.

screwed into position (Illus. 453). As the cup is tightened, grease is compressed and forced into the arbor bearings. Once the grease is compressed slightly, the cup is turned about *one* turn per *ten* hours of operation. Avoid overlubrication. The extra grease turns into liquid when the saw becomes warm, and is slung by the arbor onto the saw and work.

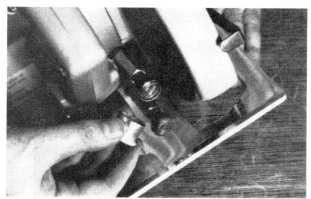

Illus. 453. Replace and tighten the grease cup. Wipe all fittings clean. This will reduce the accumulation of sawdust.

Some portable circular saws require partial disassembly when you lubricate the gears, which is done every two to six months, depending on the type of use. Hard use and adverse conditions require more frequent oil changes than shop use under ideal conditions.

Worm-drive saws have a crankcase that is filled with oil (Illus. 454). The worm gears turn in the crankcase and are constantly bathed in oil. Check the oil level periodically. Check it more frequently during heavy use. Change the oil according to the manufacturer's instructions. Make sure that the correct oil is used. An incorrect lubricant may do more harm than good. Avoid overfilling a worm-drive crankcase. This is harmful; the oil could leak into the electric motor compartment.

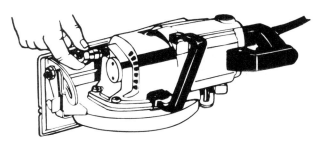

Illus. 454. Check the oil on the worm-drive saw by laying it on its side. Avoid overfilling the crankcase. This could damage the electrical components.

Many saw manufacturers suggest that you wipe a little oil or grease on the inner arbor washer when you change blades. This increases the blade's ability to slip when it becomes bound in the saw kerf. This type of lubrication is not required on a worm-drive saw.

Electrical Maintenance

Changing Brushes The carbon brushes on many portable circular saws should be changed when they become less than ¼ inch long (Illus. 455). The brushes

Illus. 455. Check the brushes periodically for wear. Replace them when they are ¼ inch long. Always check the brushes with the power disconnected.

provide electrical contact at the commutator, so disconnect the saw for this operation. On some saws, you will have to remove the outer housing for access to the brush caps. Other saws have access on the outside of the housing.

Remove both brushes and replace them as a pair if necessary. Inspect the commutator at this time for wear. If it is worn or damaged, the saw may have to be reconditioned by the manufacturer. If the commutator is in good condition, replace the brushes.

New brushes will cause considerable electrical sparking until they wear to the arc of the commutator. This is normal. As the saw is used, the spark will die down to a small bluish-white spark.

Switch and Cord Replacement No one wants to admit that he has sawed the supply cord on the saw in half, but it does happen. Supply cords are also damaged in the field through a number of causes. Repair small nicks in the outer insulation with electrical tape as soon as they occur. When bare wire(s) is exposed, replace the cord. Supply cords are sold ready-made for most saws. Disconnect the saw and disassemble the handle. Disconnect the old cord and position the strain reliever. Hook up the wires correctly. It is a good idea to mark the wires before you disconnect them. Today, most wires are color-coded, so this is not usually a problem.

The switch is the most frequently used part of any portable power tool. It is also subject to failure. Most switches are available as a spare part (Illus. 456). Many supply houses keep switches in stock. Disconnect the saw and disassemble the handle. Make a wiring diagram. Remove the old switch and replace it. Use the wiring diagram to make sure that the wires are correctly connected. Reassemble the saw and test the switch.

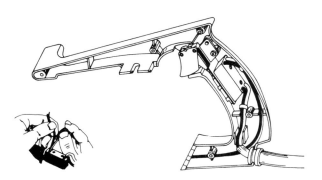

Illus. 456. Switches are available that can be used as spare parts. Replace the switch at the first sign of trouble.

Buying a Portable Circular Saw

Buying a portable circular saw can be divided into two distinct parts: selection criteria and evaluation criteria. Selection criteria involves a comparison of your needs to the types of saws and features available. Evaluation criteria concerns those factors you would use to decide which particular saw to buy. Over a dozen portable circular saws may meet your selection criteria. Evaluation criteria help you decide which particular saw is the best for the job or offers the most value for the intended purpose.

Selection Criteria The best way to determine your needs is through a series of questions. First, what type of wood will you be cutting? Consider the thickest, hardest stock that you will be cutting. This will help you decide which saw-blade diameter and power requirements will be needed. If you plan to do a lot of heavy ripping, extra power may be required. This suggests a worm-drive type saw and a splitter to help minimize the chance of kickback.

The moisture content of the wood you are cutting will also affect power needs. The higher the moisture content, the heavier the wood chips. More power is needed to remove these chips from the kerf. Hardwood chips are generally harder than softwood chips, but there are some exceptions. For example, basswood (a hardwood) is much lighter than Douglas fir and Southern yellow pine (both softwoods).

Second, where will the saw be used? If the saw is to be used on a rooftop, its weight might be a consideration. Rooftop use may suggest a cordless saw to eliminate the tripping hazard of a power cord. Shop use may suggest a saw of lesser power since most heavy cuts will be performed with stationary equipment. Damp conditions may suggest the need for a double-insulated or cordless saw.

A saw used in the field for framing will have different requirements than one used for trimming. If the saw will be used to cut the tar-laden exterior sheathing, you can select an inexpensive saw for this purpose only. This keeps the better saws clean and ready for jobs requiring greater power and accuracy. If the saw will be used for masonry or concrete cutting, the appropriate accessories should be available. These accessories may include a water coolant device. The water improves cutting and keeps the dust down. Air-driven saws are commonly used with concrete work.

From the selection criteria, you can determine the following:

Blade Diameter _____ Drive type _____
Type of Power: Battery _____ Air _____
　　　　　　　　Electric _____
　　　　　　　　Internal Combustion _____
Horsepower Requirements _____
Quality of Saw Needed: Low _____ Med _____
　　　　　　　　　　　　High _____

These factors can be used to identify a group of portable circular saws that can be used for the job.

Evaluation Criteria After you have identified a group of portable circular saws capable of doing the job, you will want to purchase the one that offers the greatest efficiency and value for its cost. The decision to purchase a particular saw may be based on a single criterion such as it is the cheapest 7¼-inch saw available, or it may be based on a number of criteria. These criteria are listed and discussed in Table 2.

Safety The single most important evaluation criterion for any portable circular saw is safety. Check the guard for its precision and its return. The guard should not wobble from side to side, and it should return quickly after a cut. Make sure that the saw is grounded properly or double-insulated to avoid shock. The noise level is also important—it's your

PORTABLE CIRCULAR SAW EVALUATION CRITERIA

Saw Brand and Model _____

Safety Check where appropriate.
Guard   _____ Substantial   _____ Quick Return   _____ Electronic Brake
Electrical   _____ Grounded   _____ Double-Insulated
Trigger Switch   _____ Safety Lockout   _____ Positioned Correctly
Noise Level   How quietly does the saw run? _____
Splitter   Saw has a splitter for safe ripping _____ Yes _____ No
Dust   Saw has a connection for dust collection _____ Yes _____ No

Power
Adequate Horsepower   _____ Yes   _____ No
Amperage   _____
RPM's   _____
Blade Diameter   _____

Use Factors / Maintenance / Longevity
Parallel _____ or Pivot Base _____   Plunging Capability _____
Worm Drive _____ or Direct Drive _____   Right Hand _____   Left Hand _____
Base Thickness _____   Construction Material _____
Weight _____   Balance of Saw _____
Overall Size _____
Handle Position(s) _____
Accessories   _____ Provided   _____ Available
Cord   _____ Location   Size _____
Bearing Type   _____ Bushing   Needle _____   Ball _____
Ease of Maintenance   Lubrication _____
  Changing Blade _____   Replacing Brushes _____

Table 2. Study the criteria for evaluating saws listed here and use them to appraise any portable circular saw you are thinking about buying.

hearing that's at stake. All other factors being equal, you should purchase the saw which runs most quietly.

The trigger switch and the splitter are other factors that must be considered. A good switch has a lockout that must be tripped before the saw will start. This eliminates accidental starting. The splitter is essential if you expect to rip frequently with your portable circular saw.

Power A saw with inadequate power will do the job poorly and wear out prematurely. Use the horsepower as an indicator of power, and then consider amperage. This relationship is discussed in Chapter 4 (page 46). Look at the power cord on the saw. It should be large enough to carry the amperage. A small cord may act as an electrical resister during heavy cuts.

The saw's rpm rating is also important. The higher the rpm's, the smoother the saw will cut using any given blade. This is because the rim speed increases with the rpm rating. Electronic saws will maintain constant rpm's under any load; this makes them desirable for heavy cuts. The rpm level on a cordless saw diminishes as the battery charge drops. Pneumatic saws maintain constant rpm's regardless of load. A complete discussion of power factors can be found in Chapter 4 (pages 48–50).

Use Factors Factors in this category relate to how the saw will be used and how easy it is to maintain. Basic decisions such as whether to get a saw with a parallel or pivot base, worm or direct drive, or right-hand or left-hand drive are discussed on pages 49–56.

The saw base is a basic component and is very important to the overall evaluation of a saw. It is frequently overlooked, however. The saw base should be substantial. A thin saw base will twist or warp easily if the saw is dropped or otherwise abused. Bases made from nonferrous materials do not rust and require no paint. Paint can rub off on the work. Aluminum is soft, and may dent easily, so it must be kept smooth. A file will remove any nicks.

The way the base attaches to the saw is also important. The more substantial the mounting, the less likely the base is to twist or warp.

Consider the weight, balance, and size of the saw in your evaluation. If you will be working in tight quarters, a small saw may be desirable. Larger saws are easier to balance and control in most cases. The saw must feel balanced when you use it. Heavier saws may be easy to use if balanced properly. Extra weight may mean that the saw has a more substantial construction, so do not consider a heavy saw undesirable solely because of its weight.

Handle positions can also affect balance and control. Most saws have both handles on the motor unit, but some have the second handle positioned elsewhere. A second handle on the base actually enables you to steer the saw as you cut.

Bearing type is also an important consideration in saw selection. A bronze bushing or sleeve will not hold up as well as needle or ball bearings. Eventually, the arbor will wobble due to wear and thrust stresses. A wobbling blade causes tear-out problems. The ideal saw has ball bearings at all wear points, but needle bearings also work quite well.

Maintenance factors such as lubrication, changing blades, and replacing brushes, as well as the availability of accessories, are important evaluation factors. The importance of these factors will vary among the individual portable circular saw users.

The saws shown in Illus. 457–460 have features that you must consider or you may desire when you select a portable circular saw.

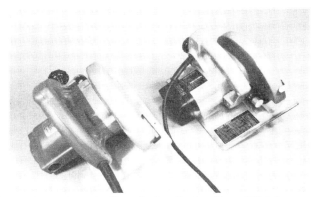

Illus. 457 (above left). Plunging saws have no lower guard. This one is equipped with a dust collector. Illus. 458 (above right). The point where the cord connects to the saw is a matter of concern to some woodworkers.

Illus. 459 (above left). A substantial base plate or shoe is important to most woodworkers. This base is made of aluminum and has a saw-guide notch cut in it. Illus. 460 (above right). Many saws now have provisions for carrying the wrench used to change the blade. This could be a desirable feature.

Mitring Machines

8

Plunging or Chop-Stroke Mitring Machines

Chop-stroke mitring machines are used chiefly for trim installation at the job site. Some of these machines are also used in the shop for picture-frame work and cabinetmaking. The blade and motor pivot off a point behind the fence. The motor is similar to the one used on a portable circular saw, except that it usually turns faster and handles a larger blade.

A 7¼-inch portable circular saw will turn about 4,000 rpm's. A 10-inch mitring machine will turn at 4,000–5,000 rpm's. The higher rpm's and larger blade diameter means faster peripheral or rim speed. This produces smoother cuts at the same feed speed.

Most mitring machines have a direct-gear drive motor, while some use a cog belt to generate the high rpm's. Chop-stroke mitring machines use blades from about 8–16 inches in diameter.

The mitring machine is designed to accommodate a blade of specific diameter. Do not use a blade larger or smaller than specified in the owner's manual. When selecting a blade, make certain that it is rated for the high rpm's of a mitring machine. Low-speed blades can break down if they are used at rpm levels above their rated capacities.

Classifications

Chop-stroke mitring machines can be classified as single-mitre or compound-mitre machines. A single-mitre machine has no provision for blade tilt (Illus. 461). The compound-mitre machine allows for blade tilt (Illus. 462). It can cut compound mitres in a flat plane. A single-mitre machine can cut compound mitres if the stock is tilted (Illus. 463).

The table on the mitring machine can be made of wood or metal. Newer designs favor metal tables that turn with the blade. Wooden tables are destroyed when the blade cuts out the center of the table (Illus. 464). They are useful for attaching jigs or stops; some woodworkers prefer them for this reason. Metal tables can be covered with a wooden auxiliary table, if de-

Illus. 461. The single or simple-mitre machine has no provision for blade tilt.

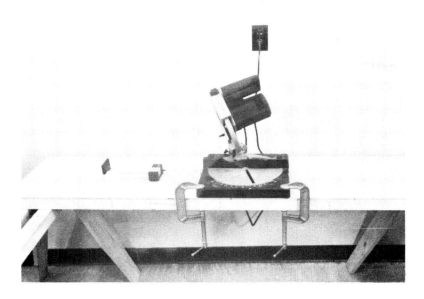

Illus. 462. The compound-mitre machine has a blade-tilting provision, which means compound mitres can be cut easily.

Illus. 463. When stock is tilted, a compound mitre can be cut on a simple mitring machine.

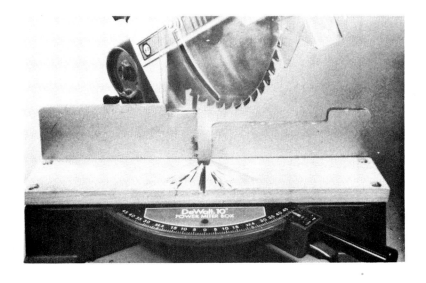

Illus. 464. A wooden table is expected to be abused. As different mitre cuts are made, the blade cuts out the center.

sired. This may require that holes be drilled in the metal table. Drilling should be done carefully.

All mitring machines have mounting holes in the feet, with which they can be anchored to a sawhorse or other sawing accessory (Illus. 465). Accessory devices will be discussed later in this chapter.

Chop-stroke mitring machines have a stop mechanism and clamping mechanism that control blade turning and positioning. Single-mitre machines have a handle that twists for clamping at the desired setting (Illus. 466). There is also a push-button stop that allows the blade to lock at common settings such as 22½, 45, and 90 degrees (Illus. 467–469). The stop can be above or below the clamp. The clamp is used with the stop.

The compound mitre saw has a mitre latch on the side of the machine that acts as a stop. Clamping is accomplished by using the two mitre clamp knobs that are located on the fence (Illus. 470). Clamp all mechanisms securely before making a cut.

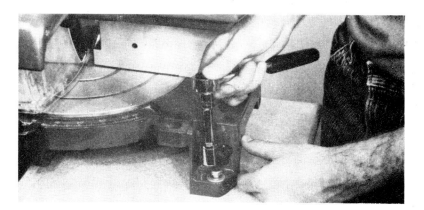

Illus. 465. The holes in the feet of the motorized mitre box allow it to be clamped to a workbench, sawhorse, or other stationary object.

Illus. 466. You can lock the cutting angle on most mitring machines by turning the handle. This provides a clamping force that holds the setting.

Illus. 467. This push-button stop is controlled by your thumb. It has positive index stops at common mitre settings.

Illus. 468. This stop is lifted with your fingers. It, too, has positive index stops.

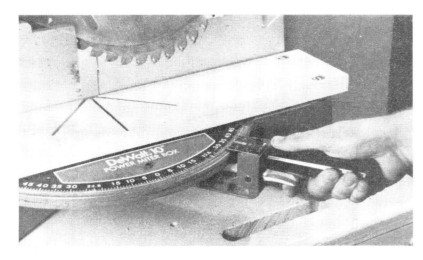

Illus. 469. This handle also has a stop that is lifted with your fingers.

Illus. 470. The knobs on the fence control the mitre setting of the blade. To hold the desired setting, tighten them.

Parts

All chop-stroke mitring machines have a table and fence. They are usually attached to the base or moulded as one piece. The rotating table turns in the base, and is held in position with the stop and/or clamping mechanism. On some machines, a stationary wooden table is held in position with four bolts.

The rotating table has a slot to accommodate the blade on its chop stroke. Newer machines have a plastic insert that fits the slot. This insert is cut by the blade, but provides support for the work adjacent to the blade (Illus. 471). This support minimizes grain tear-out as the cut is made.

The arm on a mitring machine is locked in the down position for transportation. On most machines, a pin or lever locks the arm down (Illus. 472–474). Older designs used a chain to hold the arm down, but this is no longer common. Some machines have a handle mounted on the motor. This makes the unit easier to transport from job to job (Illus. 475).

Accessories and Features

One common accessory offered with most mitring machines is an extension table and stop (Illus. 476–478). The extension table supports the work while it is being cut, and the stop controls the length of the cut (Illus. 479). This keeps all parts uniform when quality cuts are required.

Clamping mechanisms are offered as an accessory

Illus. 471. The plastic insert backs up the workpiece as the blade cuts through the work. This improves the cut by minimizing vibration and tear-out.

Illus. 472. This sliding pin locks the arm in the down position for travel.

Illus. 473. This compound-mitre saw also has a sliding pin that holds the arm in the down position.

Illus. 474. This motorized mitre box has a lever that flips over and holds the arm down for storage or transportation.

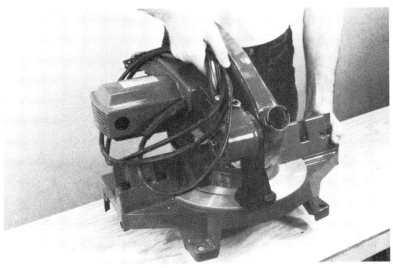

Illus. 475. This motorized mitre box has a handle mounted to the saw. The handle makes it easier to lift and transport.

Illus. 476. This extension table and stop help control longer pieces. The stop can be used to cut parts to a uniform length.

Illus. 477. The extension table is held in position with a threaded fastener.

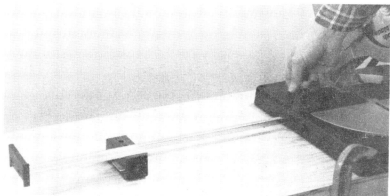

Illus. 478. The extension table on the compound-cutting mitre saw is also held in position by a threaded fastener. It also has a stop for cutting parts to uniform length.

Illus. 479. The stop controls the length of the part. Set the stop by using a tooth that points towards the stop.

with some mitring machines (Illus. 480). The clamps fit into the base casting and hold the work while it is being sawn (Illus. 481). In some cases, if the blade is turned to the left, the clamp must be used on the right side of the mitring machine. The upper arm of the mitring machine contacts the clamp when it is used on the same side. Conventional clamps can also be used to hold stock when cutting (Illus. 482).

The saw blade is protected by an upper and a lower guard. The upper guard is usually permanently attached over the upper half of the blade. It is made of stamped or cast metal. The lower guard is like the telescoping guard on the portable circular saw. It lifts up into the upper guard during the chop cut (Illus. 483 and 484). The lower guard is usually made of clear plastic; this improves vision and cutting accuracy. It is spring-loaded; it returns after every cut. It should never be wedged or taped in the up position.

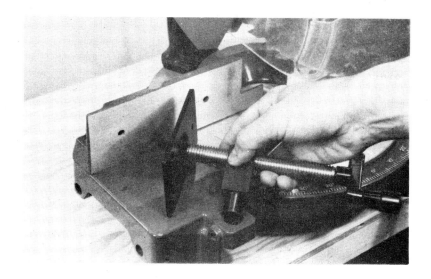

Illus. 480. Clamping mechanisms are frequently offered as an accessory to the motorized mitre box. They fit openings in the base of the saw.

Illus. 481. The clamping device holds small parts for sawing. This keeps your hand clear of the blade while the cut is being made.

Illus. 482. Conventional clamps can also be used to hold stock while it is being cut.

 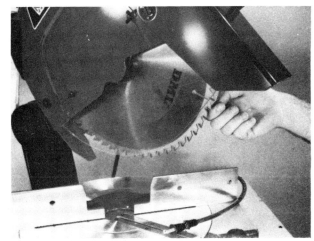

Illus. 483 (above left). The lower guard on the compound-mitre saw lifts mechanically as the arm goes down. Illus. 484 (above right). The clear plastic guard on the motorized mitre box is lifted by the workpiece.

Many mitring machines have a braking system as a safety feature. The braking system can be mechanical or electronic. The mechanical brake is actuated with a plunger-type button; when the button is hit, the mechanical brake is applied to the blade or drive mechanism. Electronic braking systems reverse the current through the motor, causing the blade to stop quickly. Electronic brakes tend to cause less wear and damage to the mitring machine.

Mitring machines have a variety of handles that are used to control the plunging action (Illus. 485–488). Many woodworkers favor the flat or straight handle because it makes it easier to plunge the blade. Some machines have a D-style handle or other design. These designs do not always make the plunging action as easy as the flat handles does.

Illus. 486. This handle is vertical. It is easy to grip, but not as easy to plunge.

Illus. 485. The handle on this motorized mitre box is horizontal and round. It is easy to grip.

Illus. 487. This horizontal handle is parallel to the fence. It is comfortable and easy to plunge.

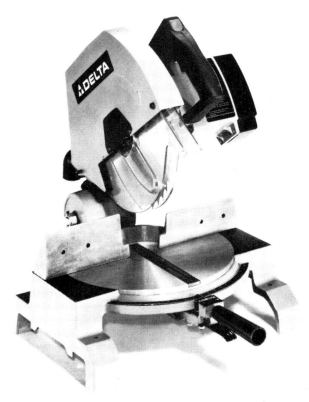

Illus. 488. This vertical handle is also easy to grip and plunge.

A desirable feature offered as an accessory for most plunging mitre machines is the dust bag or catcher. The high rpm level of the mitring machine causes it to send fine dust a long distance. The bag attaches to the dust chute and collects most of the sawdust produced by the saw (Illus. 489). The dust bag can reduce the mess generated when you use the chop saw. In many

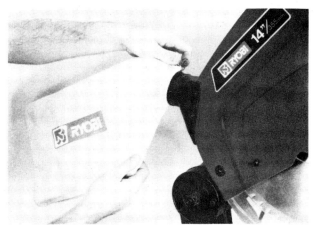

Illus. 489. A cloth dust bag is offered with many motorized mitre boxes. It contains most of the dust generated by the saw.

cases, a dust collector (Illus. 490) can be attached to the spout on which the dust bag mounts. The vacuum removes most of the dust from the air and contains it at the collector.

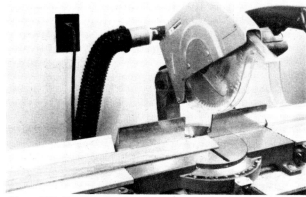

Illus. 490. Dust-collection hoses can also be fitted to the motorized mitre box. The dust-collection system can catch all dust generated by the saw.

General Safety Guidelines

Before operating any plunge- or chop-style mitring machine, read, understand, and observe the following precautions and those listed in the owner's manual. Review the owner's manual and the general safety precautions periodically as you work. This will help you remain alert to the common causes of mishaps.

1. Make sure that the mitre saw is mounted securely to a work surface or bench. The saw must remain stable during use.

2. Make sure that the mitre saw is grounded properly. Use a ground-fault circuit interrupter when working in the field. Many of today's motorized mitre boxes are double-insulated to guard against electrical shock.

3. Make sure that the upper guard is mounted securely, and that the lower guard is retracting properly (Illus. 491). Never wedge the lower guard into the up position.

4. Make all adjustments and blade changes with the power disconnected (Illus. 492).

5. Protect yourself with hearing protectors and protective glasses or goggles.

6. Make sure that fences, stops, and clamps are adjusted properly and locked securely before cutting. Check them periodically when doing production work.

7. Make sure that the stock is butted or clamped securely to the table and fence before making any cuts (Illus. 493).

Illus. 491. Check the lower guard periodically to be sure that it moves freely. A lazy guard can cause problems and create a hazardous condition.

Illus. 492. Blade changes and other adjustments should be made with the power disconnected.

Illus. 493. Make sure that the stock is butted or clamped securely to the table and fence before making any cuts. This mitre box fits over dimensional stock to make table extensions.

8. Keep your hands clear of the blade's path when using any mitring machine. Keep a 6-inch margin of safety from the blade's path for all operations.

9. Never cross your hands when using any mitring machine. This could put one hand in or near the blade's path. A serious injury could result.

10. Keep yourself balanced while cutting. Avoid overreaching. This could lead to a mishap.

11. Allow the blade to come up to full speed before starting the cut. Keep the cord clear of the blade's path.

12. Do not force the saw. Let it cut at a steady rate.

13. Shut off the saw after the cut. When the blade stops, lift it out of the cut. Never lift a spinning blade out of the cut. It could catch on the cutoff and throw it. It could also bend a tooth on the blade.

14. Make sure that the brake is working properly before using any chop-stroke saw. The blade should stop quickly after you have released the switch.

15. Be sure to clamp round stock. Round stock can roll into the blade in a gear-like manner. This may damage the blade or machine and could cause an accident (Illus. 494).

16. Never attempt to hold pieces that are less than 12 inches long with your hands. Use the clamps that come with the mitre saw or use a parallel clamp to hold them to the fence. This will keep your hands clear of the blade.

17. Support long pieces with an auxiliary table or support (Illus. 495 and 496).

18. Handle warped or bowed stock properly. Make sure that it does not pinch the blade as the cut is being made (Illus. 497 and 498).

Illus. 494. Round stock should always be clamped. This keeps it from rolling into the blade when cutting begins.

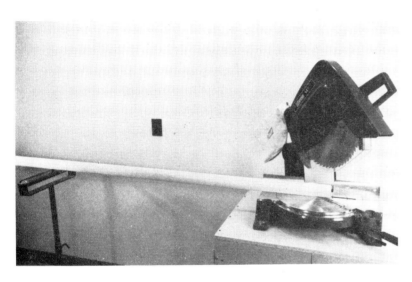

Illus. 495. This roller support helps control this piece of stock. It is difficult to hold long stock safely when it is not supported.

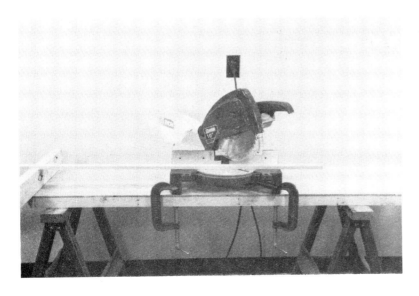

Illus. 496. In some cases, pieces of scrap can be used for stock support. Make sure that the scrap pieces are the correct height or they may push the work out of the sawing plane.

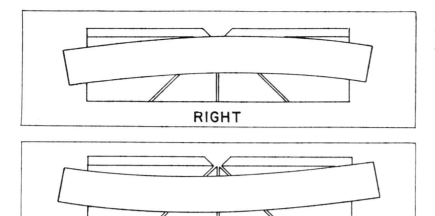

Illus. 497. Bowed pieces should be sawed with the bow butted against the fence and table.

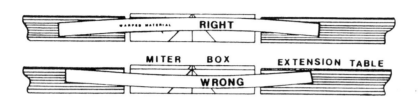

Illus. 498. When using extension tables, you will find that even a slight bow can be a problem. Always inspect your stock.

Common Operations

The common operations performed by the chop-stroke motorized mitre box include crosscutting, simple-mitring, and compound-mitring. Each of these operations is discussed under its own heading.

Crosscutting Mitre boxes are frequently used to crosscut parts. In some cases, only one part is cut; in other cases, a large number of parts of uniform length must be cut. Set the saw at 90 degrees and lock the setting (Illus. 499). If you are crosscutting a single part, line up the layout line with the saw blade (Illus. 500). The saw blade must be on the waste side of the line. Turn on the saw and allow it to come up to full speed. Slowly pull the saw down into the work (Illus. 501). Feed the blade into the wood only as fast as it will cut (Illus. 502). Rapid plunging tends to tear the grain. This can tax the machine and waste material. Let the blade come to a complete stop before raising it from the cut. This will keep small cutoffs from being thrown.

When crosscutting several parts, use a stop to control part length. Measure from a tooth that points towards the stop (Illus. 479). Set the stop and make a test cut. Readjust the stop if necessary. Once the setting is correct, cut all the parts (Illus. 503).

In some cases, commercial tables and stops can serve as the length controls (Illus. 504–506). These accessories make cutting safer and more accurate. You can also attach a wooden table or fence to the table or fence of the mitre saw, and then nail or clamp stops to it.

A wooden auxiliary table may also be beneficial when you are cutting wide pieces. It lifts the workpiece closer to the center of the blade, where the widest possible cut can be made (Illus. 507–509).

If you are crosscutting long pieces into small parts, make sure the work is supported (Illus. 510–512). This eliminates the chance of the piece lifting during the cut. Lifting could force the operator's hand into contact with the saw blade.

Some parts have a slight angular cut on the ends. This is common on truss and rafter braces. To crosscut these parts, turn the blade to the desired angle and lock it in position. Use a stop to control stock length. Set it up the same way you would for any other crosscut. Test the setup before cutting your parts.

Cutting Simple Mitres When you are cutting simple mitres, the blade has to be turned to the correct mitre

Illus. 499. For crosscutting, set and lock the blade at 90 degrees.

Illus. 500. Line up the blade with the layout line. Let the saw come up to full speed and begin the cut.

Illus. 501. Feed the blade into the wood only as quickly as it will cut. Let the blade come to a complete stop at the bottom of the cut.

Illus. 502. Raise the blade from the work after it has come to a complete stop.

Illus. 503. With the stop, several parts can be cut to exact length. Make the setup carefully for greatest accuracy.

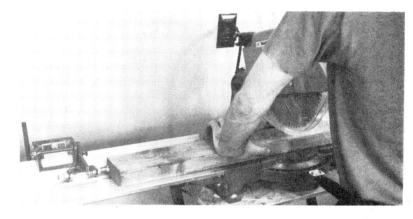

Illus. 504. This Flipstop™ controls the length of the stock. It clamps to the fence. Minor adjustments can be made by adjusting the flathead machine screw.

Illus. 505. The Rack tubing system has a support/stop combination.

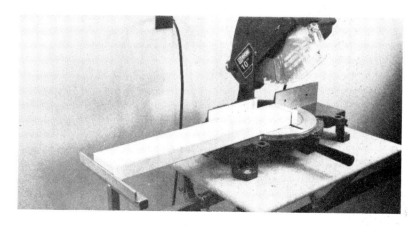

Illus. 506. The stock is supported by the tubing after the cut is made.

Illus. 507. This piece of stock was too wide for a complete crosscut.

Illus. 508. An auxiliary table was added to the motorized mitre box table.

Illus. 509. This cut was completed by raising the work up to the widest part of the blade.

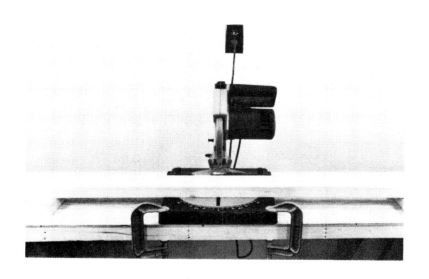

Illus. 510. Scraps are supporting the stock on this compound-mitre saw.

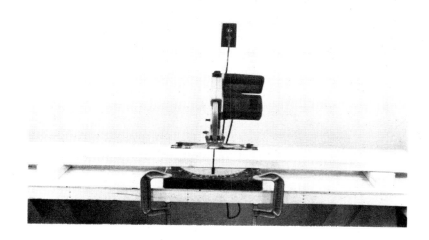

Illus. 511. The cut is made safely without the pieces lifting into the blade.

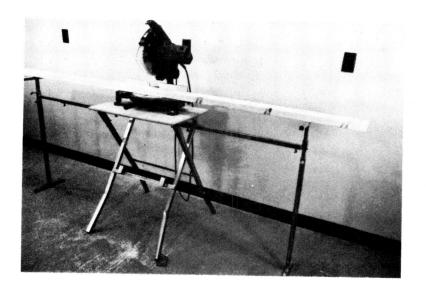

Illus. 512. This tubing system also provides support for long pieces.

angle. In most cases, this may be 22.5 or 45 degrees. Most machines have stops at these settings to make the setup easier and more accurate. Some mitres will not be as easy to set up. First make these cuts in scrap to be sure that the setting is correct. Once the setting is correct, lock the saw at that setting.

When using compound-mitring machines, you also have to turn the blade to the desired angle (Illus. 513–516). Release the clamp knobs on the fence and depress the stop button on the side of the machine. Turn the blade to the desired angle and release the stop button. The stop button will engage at common settings such as 90 degrees, 45 degrees, and 22.5 degrees. At other settings, the clamp knobs on the fence hold the blade at the desired setting. The clamp knobs are used in conjunction with the stop button. They make the setting more rigid.

When using simple-mitring machines (Illus. 517), line the stock up with the blade (Illus. 518). Turn the blade on and, once it is at full speed, make the cut (Illus. 519). After completing the cut, turn the saw off and lift the motor unit to the top of its stroke (Illus. 520). Use a stop to control the length of mitred parts (Illus. 521 and 522). This stop can be on an auxiliary table or on a table extension. You can also make simple mitre cuts with the stock vertically oriented. This is known as an end mitre (Illus. 523–526).

On compound-mitring machines, line the stock up

Illus. 513. Turn the turntable on the compound-mitre saw to the desired angle and clamp it.

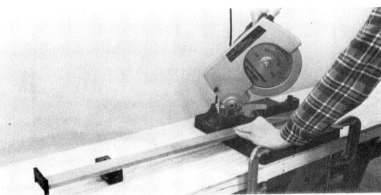

Illus. 514. Make the mitre cut using the stop to control the length of the parts.

Illus. 515. Lift the blade only after the blade has come to a complete stop.

Illus. 516. The cut is smooth, with no tear-out. The cutoff did not move because the blade was stopped before it was lifted.

Illus. 517. Set the mitre using the stop and clamping mechanism. Clamp the setting securely.

Illus. 518. Line the stock up with the blade and butt the stock securely to the fence and table.

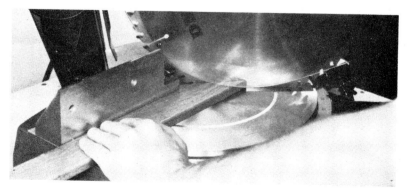

Illus. 519. When the blade comes up to full speed, push it into the work. Shut off the saw when the cut is complete.

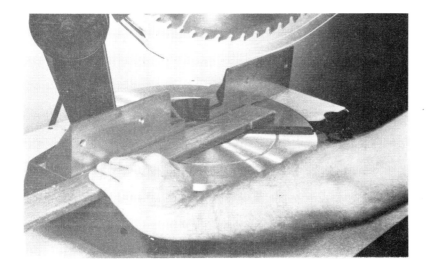

Illus. 520. When the blade stops, raise it out of the workpiece. This keeps the cutoff from being ejected.

Illus. 521. This stop controls the length of the mitred part. This setup is used for production jobs.

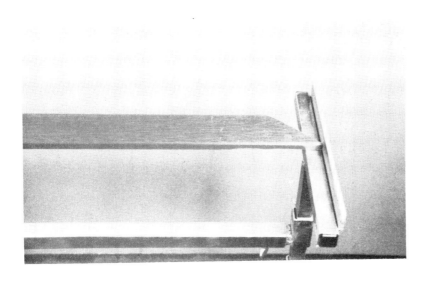

Illus. 522. The mitre butts against the stop to control the length of the part. Work carefully so that the mitre point is not damaged.

Illus. 523 (above left). Position the stock for end-mitring. Clamp the blade adjustment securely. Illus. 524 (above right). Begin the cut when the blade comes up to full speed.

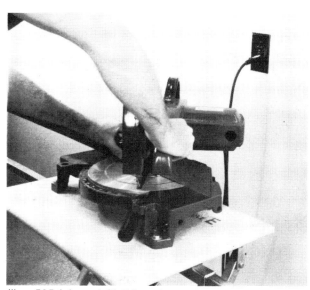 

Illus. 525 (above left). When the cut is complete, shut off the saw. Illus. 526 (above right). Lift the saw out of the cut when the blade comes to a complete stop.

with the blade. Release the lower blade guard to allow it to retract. Lift the motor unit to its full height, which restores the lower guard to its original position. Turn on the motor and let it come up to full speed. Depress the lower guard and begin the cutting stroke. When you have completed the cut, shut off the saw. Lift the motor unit to the top of its stroke. You can use a stop to control the length of mitred parts. The stop on the machine can be used for shorter parts. Longer parts will require a commercial device.

When installing base or ceiling trim, you will find that many trim pieces are too short to extend from one corner to the other. In this case, use a splice or scarf joint (Illus. 527 and 528). The splice is usually made at a framing stud, so both parts can be nailed to the wall securely. The splice or scarf joint is just two mating 45-degree angle cuts that overlap at a stud. They are cut the same way as any other end mitre.

Illus. 527. Two complementary end mitres make a splice or scarf joint. This joint is used when moulding cannot extend the full length of the work space.

Illus. 528. When the scarf joint is pressed together, it is almost invisible.

Cutting Compound Mitres A compound mitre is a mitre with two angles instead of one. Parts that are tilted, such as cove moulding and frames, often require compound-mitre cuts. These cuts can be made on a simple- or compound-mitring machine.

Compound mitres are cut on the compound-mitring machine in the flat plane. To cut a compound mitre on a simple-mitring machine, place a triangular block against the table and the fence; it can be held in place with double-faced tape. This block tilts the work to the desired angle and allows the simple-mitring machine to cut a compound mitre.

Once the block is attached to the table and fence of the simple-mitring machine (Illus. 529), turn the blade to the desired angle and lock it in position. The work touches the fence, table, and triangular block. Then cut the mitre the same way you would a simple mitre (Illus. 530).

It is important that the tilt of the work and the triangular block be accurate. Any deviation in the angle can affect the fit of the mitre. Minor adjustments

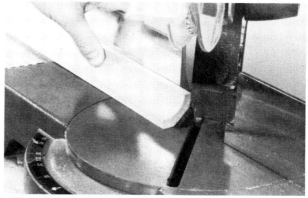

Illus. 529. A triangular block is supporting the work so that a compound-mitre cut can be made.

Illus. 530. The blade cuts the work and the triangular work support.

can be made by building up an end of the triangular block with masking tape. This changes the work tilt a few degrees.

Wooden fences and tables can be attached to mitring machines with a metal table (Illus. 531). You can use a cleat to hold the work at the desired tilt (Illus. 532). Tack the cleat to the wooden table. The base of the work butts against the cleat, while the opposite edge butts against the fence. Make sure that the nails that hold the cleat in position are not in the blade's path. Cut the mitres the way you would cut a simple mitre (Illus. 533–536). If you are using a mitring machine with wooden tables, just nail a wooden cleat to the table to obtain the same results.

Compound-mitring machines are set up somewhat differently. The work lies in the flat plane (Illus. 537–540), which means that the blade must be both turned and tilted. The incline of the work determines the blade angle and the blade tilt. Use the angles on the chart in Chapter 9 (page 239) to set up the saw. Check your setups on scrap to be sure they are correct.

Compound mitres require careful setup because the blade is both turned and tilted. Some readjustment is usually necessary to get a perfect mitre. Do not rush the setup. If you do, the work will be spoiled. Deep pictures frames, wastebaskets, and planters are all assembled with compound mitres.

Organizing compound mitre cuts is difficult. For picture frames, proceed as follows:

Right mitre: Use the fence to the right of blade.

Illus. 531A (above left). These two pieces are glued and clamped together to make an auxiliary table and fence for the motorized mitre box. Illus. 531B (above right). The auxiliary table and fence unit is secured to the fence of this motorized mitre box.

Illus. 532. A cleat can be tacked to the table or fence to hold stock at the desired angle. This picture-frame stock is supported by the glass rabbet.

Illus. 533. You can cut mitres in the conventional manner when you use auxiliary fences and tables.

Illus. 534. Allow the blade to come up to full speed before starting the cut.

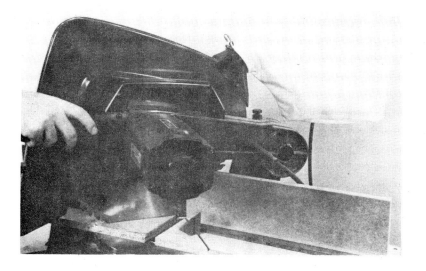

Illus. 535. Shut the saw off when the cut is complete.

Illus. 536. Lift up the blade after it comes to a complete stop.

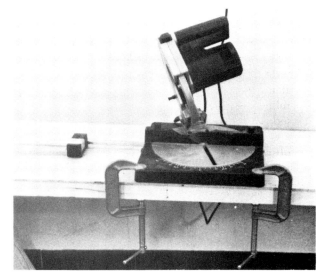

Illus. 537 (above left). Set one angle of the compound mitre by tilting the blade. Illus. 538 (above right). Set the other angle by turning the blade. Clamp the setting using the knobs on the fence.

Illus. 539. Adjust the stop if desired. This will control the length of the stock.

Illus. 540. Make the cut the same way you would make any other mitre. Let the blade come to a complete stop before lifting it from the work.

The rabbet face should be away from fence and the exposed face down.

Left mitre: Use the fence to the left of the blade. The rabbet face should be away from fence and the exposed face up. Always work on scrap to improve the setup. Compound-mitre cuts can waste a lot of stock if the setup is incorrect.

Remember to use the length gauge when making frame parts. Set the length gauge when cutting the second mitre; this will keep the parts' lengths uniform and make frame assembly easier. Remember to use the glass rabbet to set up the parts' lengths. The rabbet must be long enough to accommodate the glass or framed object.

Cutting Crown Moulding

The chop-stroke compound-mitring machine works well for cutting crown moulding. Since the two angles of the crown moulding are not equal, the saw settings have to be set at odd angles. Use the chart in Illus. 541 to set the machine. These angles are difficult to set and should be tested on scrap (Illus. 542 and 543). Remember, plastered, taped, or panelled walls rarely have a perfect 90-degree corner.

Scarf joints will be needed whenever lengths will not span from corner to corner. These are cut the same way as inside and outside corners. If you use auxiliary tables and fences, large mitre boxes can also make this crown moulding cut (Illus. 544–549).

Making a Cope Joint

On many jobs, inside corner mouldings and door stop mouldings are not fitted with mitre joints. Instead, a cope joint is used. A cope joint looks like a mitre, except wood shrinkage does not cause the joint to open up.

A cope joint is cut by hand. Make a 45-degree angle mitre cut on the work first. This provides a layout line to guide the coping saw (Illus. 550). Many carpenters make provisions for cutting cope joints at the mitring bench.

The work is usually clamped to the bench. Install a medium blade in the coping saw and begin the cut (Illus. 551). Take short strokes until you start the cut. When a tight turn is necessary, twist the blade in the saw frame (Illus. 552). As you pull the blade into the work, it turns its path slightly. Continue the cut with the blade straight.

In some cases, you will back-cut or relieve the cope joint (Illus. 553); this means that the joint is only

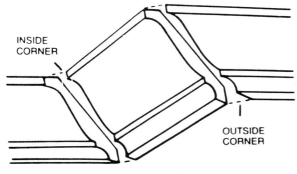

Illus. 541. Use this drawing and chart to learn how to cut compound mitres on crown moulding.

BEVEL POST SETTING	TYPE OF CUT
33.85°	LEFT SIDE, INSIDE CORNER: 1. Top of molding against fence 2. Miter table set right 31.62° 3. Save left end of cut
33.85°	RIGHT SIDE, INSIDE CORNER: 1. Bottom of molding against fence 2. Miter tables set left 31.62° 3. Save left end of cut
33.65°	LEFT SIDE, OUTSIDE CORNER: 1. Bottom of molding against fence 2. Miter table set left 31.62° 3. Save right side of cut
33.85°	RIGHT SIDE, OUTSIDE CORNER: 1. Top of molding against fence 2. Miter table set right 31.62° 3. Save right side of cut

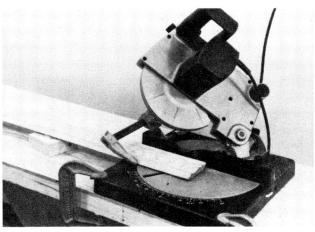 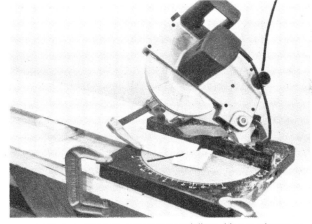

Illus. 542 (above left). Test a setup of this nature on a piece of scrap before cutting the crown moulding. Use a clamp to keep the moulding from being pulled into the blade. Illus. 543 (above right). Cut the moulding the same way you would cut other stock. Use conventional mitring practices.

Illus. 544. Crown moulding can be held at the correct angle with a cleat. Tack the cleat to the auxiliary table.

Illus. 545. Position the stock for mitring in the conventional manner.

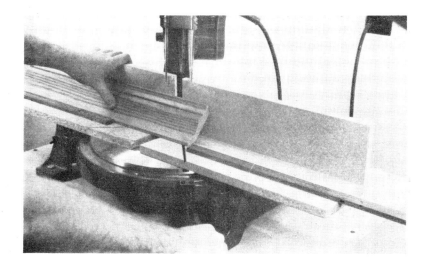

Illus. 546. Observe standard mitring practices when cutting the mitre.

Illus. 547. Raise the blade from the cut only after it has come to a complete stop.

Illus. 548. Make the mating mitre cut by turning the blade to the opposite 45-degree setting.

Illus. 549. The mating mitre cut is completed. Note the quality of the cut even when the entire workpiece is not backed by a table.

Illus. 550. The mitre angle provides a layout line for the cope cut. Clamp the work securely for cutting. A pad protects the work from damage.

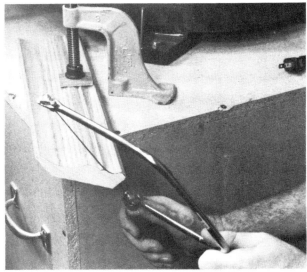 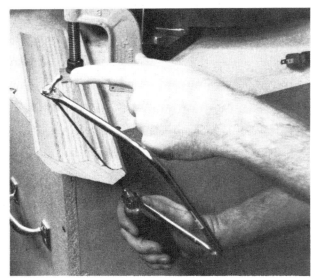

Illus. 551 (above left). Use a coping saw to make the cope joint. Take short strokes until you start the cut. Illus. 552 (above right). For tight turns, you can twist the blade. Work carefully with moderate force. This will minimize the chance of blade breakage.

Illus. 553. The back side of most cope cuts is relieved or "back-cut." This makes the exposed fit look better.

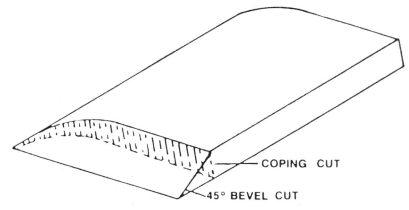

touching on the visible edge. Back-cutting provides some clearance for a good fit. Make sure that none of the back-cut area is visible (Illus. 554).

Some cope joints do not require any specialty cutting with a coping saw. The profile of the face is a 7-degree incline. When a face cut is made at 7 degrees (Illus. 555 and 556), the end forms a perfect cope with the mating part (Illus. 557).

Illus. 554. The fit between the parts should be tight. A quality cope joint makes the trimming job look much better.

Illus. 555. You can make some cope joints with a 7-degree face cut.

Illus. 556. The 7-degree cut matches the face profile of the trim stock.

Illus. 557. This cope joint fits well, yet it requires no hand sawing.

Brick-Mould Corners

Some windows can be framed on the outside with brick moulding (Illus. 558). The corner is not mitred; it is notched (Illus. 559). Cut the notch with the mitre box set at 90 degrees (Illus. 560). Stop the cut at the moulding line. Make the notch cut with a hand saw (Illus. 561). Test the fit on the mating parts (Illus. 562).

Mitre-Box Picture Frames

The motorized mitre box is used to cut the mitres on many picture frames (Illus. 563). In some cases, stock must be supported so that it can be held in the correct plane (Illus. 564). Cleats can be used to hold stock for compound mitres (Illus. 565 and 566 and Illus. 532–536).

There are commercial accessories for framing that are designed to be used with motorized mitre boxes. With the Lion™ Measuring Bench (Illus. 567), you can cut mitres with a stop. The stop controls rabbet-to-rabbet length, which is desirable for picture-framing.

The Sawhelper™ Miter-Grid™ (Illus. 568) is an accessory table that has an adhesive grid that helps in mitre measurement and cutting. The lines can be aligned with the long point, short point, or rabbet on frame stock (Illus. 569). This makes setup very easy and fast (Illus. 570 and 571).

General Trim Installation

Ceiling or base trim usually begins on a long, straight run. For ceiling or base trim, install a long piece from inside corner to inside corner with butt joints at both ends. Now, bring the trim forward out of both corners; this requires the cutting of cope joints. From this point, cut mitres for all outside corners and cope joints for all inside corners. On long runs, scarf joints will also be needed.

When carpenters trim a house or apartment, they

Illus. 558. This brick-mould corner is another corner treatment that can be cut on the motorized mitre box.

Illus. 559. Two mating notches form the brick-mould corner. There is little separation with a joint of this type. It remains tight.

Illus. 560. Stop the notch cut at the moulding line.

Illus. 561. To form the notch, cut along the moulding line with a hand saw.

Illus. 562. Test the fit between the mating parts. The moulding shape influences how the parts will be cut.

Illus. 563. Picture-framing is frequently done with a motorized mitre box. The cuts are clean and accurate.

Illus. 564. You can use a spacer to hold framing stock in a true plane for mitring.

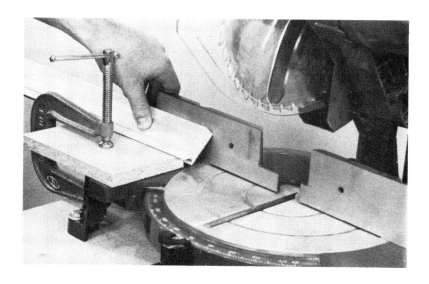

Illus. 565. You can use cleats to hold frame stock at the desired angle for mitring.

Illus. 566. Position the stock and make a conventional mitre cut.

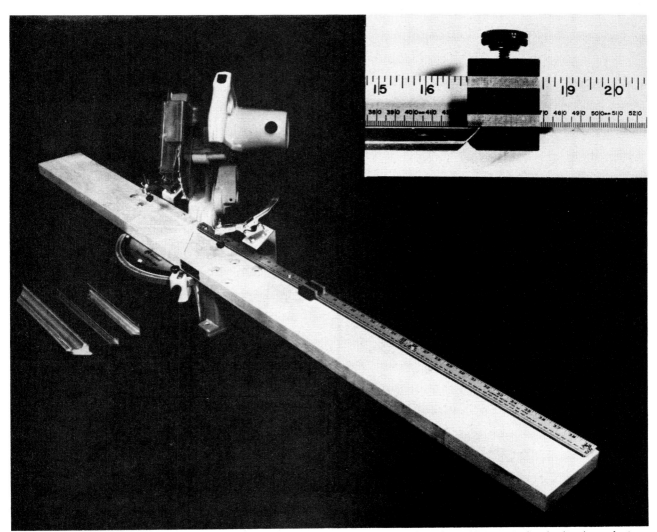

Illus. 567. To cut picture-frame stock, use the Lion Measuring Bench in conjunction with a motorized mitre box. A stop controls rabbet-to-rabbet length of the framing members.

Illus. 568. This Miter-Grid has an adhesive back. Position it and stick it to the extension table.

Illus. 569. These lines tell the frame lengths: long point, 17 inches; rabbet, 14 inches; and short point, 13½ inches.

Illus. 570. This simple-mitre picture frame was made with a motorized mitre box.

Illus. 571. The compound mitres in this frame were cut with a motorized mitre box.

may work in several rooms at the same time. One carpenter may do all the cutting, while other carpenters measure and install. Measurements such as 18¼, cope right may be "called-out"; this means that the piece is 18¼ inches long from the right-hand cope joint to the butt joint at the opposite end. When carpenters work as a team, it is amazing how fast they can install all the trim in a house.

Door and window trim differ a little in the method of installation. Some windows have a windowsill and two mitre joints in the trim that sits on the sill. Other windows have a picture-frame moulding.

For trim that originates at a windowsill, cut the two short sides to equal lengths and install them with a uniform jamb exposure or reveal. Then measure the top piece and cut it to fit the short points. Use glue and nails to hold the frame together.

Use this same procedure to trim a door. Do not mitre the stop, which holds the door when latched. Butt the head piece in position, and fit the side pieces with cope joints.

Windows on which the picture-frame moulding method is used can be installed in parts or as a frame. The parts method is similar to the trim methods just discussed. The frame method is different. For this method (Illus. 572–583), glue and assemble all the parts on a true surface. Drive nails through the corners to reinforce the joints. After the glue cures, nail the entire frame to the window. This method helps keep the mitres tight after installation.

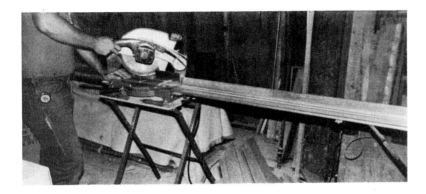

Illus. 572. Production-framing for the doors of an entire apartment house was done with this saw.

Illus. 573. A mitre stop was used to control length and protect mitre points.

Illus. 574. Glue was brushed onto the mitres to keep glue squeeze-out to a minimum.

Illus. 575. Two mitres were joined first to form an L.

Illus. 576. T-shaped nails were driven through the edges of both parts. This locked the outside corner.

Illus. 577. A staple was driven from the back side near the short point.

Illus. 578. The staple helped hold the parts together while the glue cured.

Illus. 579. The L-shaped parts were removed from the bench and joined together.

Illus. 580. The mitres were joined from the two sides of the bench with glue and nails.

Illus. 581. Frames were taped into bundles for transportation. Not a single mitre joint broke.

Illus. 582. Door frames were U-shaped. They were stacked alternately and taped, so they were easy to handle and transport.

Illus. 583. At the job, the frames were simply nailed to the opening. The amount of time we saved using this method was significant.

Determining Mitre Cuts for Odd Angles

An odd angle is the most difficult angle for determining the mitre cut. The best procedure is as follows:
1. Copy the angle using a sliding T bevel.
2. Lay that angle out on a clean scrap of sheet stock (Illus. 584).
3. Use a compass to strike an arc from the vertex of the angle. Be sure to mark both legs of the triangle (Illus. 585).
4. Extend the compass wings slightly and strike two arcs using the points made with the first arc on the legs of the angle (Illus. 586).
5. Draw a line from the vertex through the two converging arcs.
6. Copy this angle with a sliding T bevel (Illus. 587).
7. Use this angle to set the angle between the blade and fence on the chop-stroke mitring machine. Make sure that the power is disconnected for this setup.
8. Check the setup on a scrap to make sure that the fit is correct. Make any needed adjustments.

Illus. 584. To divide an odd angle, begin by laying out the angle on a clean piece of scrap.

Illus. 585. Draw an arc inside the odd angle. The size of the arc is not important.

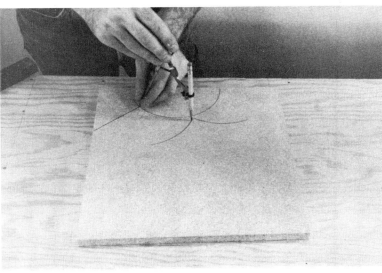

Illus. 586. Open the compass slightly and strike arcs from the intersection of the first arc and the odd angle.

Illus. 587. Connect the intersection of the arcs with the vertex of the odd angle. This angle can be used to set the motorized mitre box.

Cutting Metal and Plastic

Some plunge-type mitring machines are designed strictly for wood. Others are designed to cut metal and/or plastic (Illus. 588–590). Before using any machine to cut material other than wood, consult the owner's manual to be certain that the saw is capable of the job and that you have selected the appropriate blade.

There are machines that look similar to chop-stroke mitring machines that are used only to cut metal. These machines may be the best choice when many pieces of metal must be cut.

Accessory Tables for Mitre Boxes

Due to the popularity of the chop-stroke mitring machine, many accessory tables or supports have been developed. These tables and supports make the saw safer, more useful, and more accurate. Some tables are commercial, and others are shop-made.

Commercial Accessories Probably the simplest mix of accessories is a folding workbench or sawhorse and one or two extension rollers. When these are used, the wood will balance on the saw and an extension roller. Adjust the roller to the desired height to keep the work in a true plane.

Another commercial system is the Rack™ system. The Rack consists of a stand and related accessories (Illus. 521 and 522). All parts are made of square tubing. The stand folds up for storage. All other parts are also attached to the folded stand during storage. When used, the mitre saw is bolted to the table.

The Rack™ system has adjustable supports that extend from both sides of the mitre table. These supports hold the work at the desired height. A stiff leg is also provided for support; this leg keeps the weight of the work from toppling the unit.

Illus. 588. While this moulding looks like wood, it is actually plastic.

Illus. 589. The moulding cuts more like wood than plastic. It can be cut with a conventional blade. Other plastics require a special blade.

Illus. 590. The cut in this plastic is not as clean as it would be in solid stock.

One of the Rack supports has a stop affixed to it that allows the unit to be used for repetitive cuts. The work can be butted to the stop for length control.

Another square tubing stand is available from American Design and Engineering. It can be used with a shop-made stand (Illus. 591) or a stand marketed by American Design and Engineering (Illus. 592). The tubing attaches to the saw (Illus. 593) instead of the stand. It also has a stop that can be used to control stock length (Illus. 594).

Another system is the Sawhelper™ Ultrafence™ system from American Design and Engineering. This system offers solid tables with adjustable folding leg sets and quick-connecting couplers. The saw table unfolds and the saw is positioned on it (Illus. 595–598). The saw is bolted to a wooden table that drops into position.

The saw has couplers bolted to its ends; these couplers come with the system (Illus. 599–601). Installation templates are also furnished. The coupler on the saw fits the coupler on the table. The quick coupler is self-adjusting as it is tightened; it is accurate to $\frac{1}{100}$th of an inch after all parts are aligned. The system is easy to align, and you can set it up on your saw in a couple of hours.

The fence on the Ultrafence system also has a stop that can be used with it. It is called the Flipstop™ (Illus. 602). It slides on the fence and can be locked at any position. The work touches a flathead screw on the Flipstop. In addition, the stop flips out of the way so that you can cut other work without changing the setting of the Flipstop™.

The Rack and the Sawhelper Ultrafence systems are excellent accessories for trim carpenters, picture framers, and cabinetmakers. Both fold up for compactness and ease of transportation in apartments and condominiums. The setup of both systems is quick, accurate, and positive.

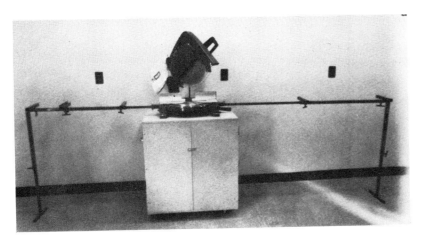

Illus. 591. The square-tubing portable work support shown here can be used with a shop-made stand.

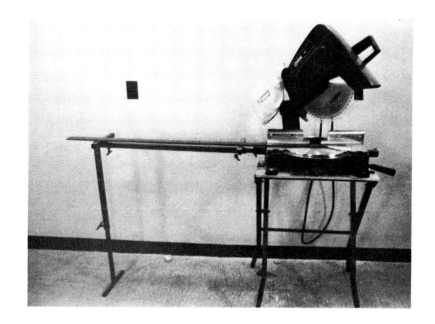

Illus. 592. A commercial stand is also available for use with this work-support system.

Illus. 593. The square tubing attaches directly to the saw. A special coupler is used to break the system down for travel.

Illus. 594. This system incorporates a stop to control the length of the stock.

Illus. 595. This folding mitre saw stand is easy to transport and store.

Illus. 596. The knob acts as a clamping device to hold the stand together.

Illus. 597. Bolt the wooden base to the saw. The cleats position it on the folding stand.

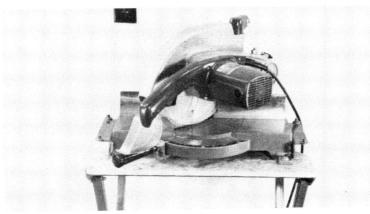

Illus. 598. Position the motorized mitre box on the folding stand.

Illus. 599. Attach and level the work-support stands.

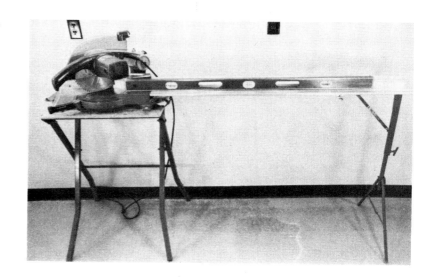

Illus. 600. You can attach one support to each side of the saw for better stock support.

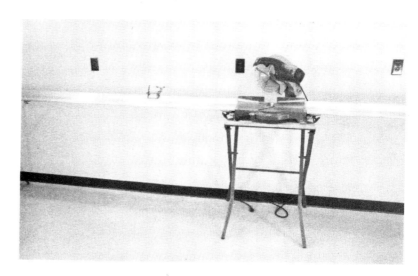

Illus. 601. The Ultrafence system set up at the job site. Note the Flipstop located on each table.

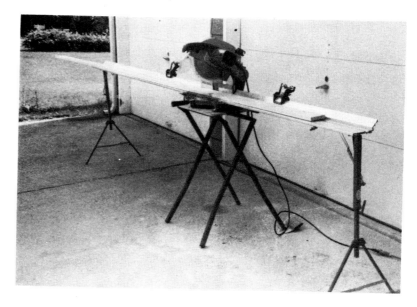

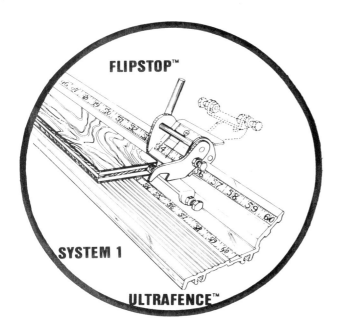

Illus. 602. The Flipstop lifts out of the way, but maintains the setting. This is useful when you are squaring ends.

Illus. 603. Always disconnect the power before changing the blade. Use the proper wrenches to change the blade. Make sure that the new blade is rated for the arbor speed of the saw.

Shop-Made Accessories Some woodworkers are not professional trim carpenters and do not want such a sophisticated system. For them, a simple box on wheels may be enough. Work supports can be sawhorses or extension rollers. In some cases, the mitre saw is simply clamped to a sawhorse.

Maintenance of Chop-Stroke Mitring Machines

There are many aspects of maintenance on chop-stroke mitring machines. These include the maintenance of the blade, table, and electrical features, and alignment of the machine. Each of these topics is discussed under the following appropriate subheadings.

Changing the Blade When the saw blade tears the wood fibres instead of cutting them, that is an indication that the blade is dull. Inspect the teeth on the blade. When the teeth on a tool-steel blade are no longer pointed, they are too dull to cut mitres effectively. If you can slide your fingernail over a carbide tip, it is too dull to cut effectively. These conditions suggest a need to change the blade. Further information about circular-saw blades can be found in Chapter 3.

Disconnect the motorized mitre box and gather the needed wrench(es) (Illus. 603) and appropriate replacement blade. The replacement blade must be the correct size (blade diameter) and must be rated for the rpm speed of the mitring machine. Never use a blade that is rated below the rpm rating of the mitring machine. A mishap is sure to result.

For best results, use a carbide-tipped crosscut blade with 40–80 teeth on a 10-inch mitring machine; use these numbers as a guideline when selecting a blade. Carbide is recommended because much of the moulding is prefinished. The finish is so hard that it will dull tool-steel blades very quickly.

On some machines, you have to remove the upper guard to change the blade. On others, an access hole on the side of the upper guard is provided.

Inspect the arbor threads to determine whether they are right hand or left hand. Loosen the arbor nut using the appropriate wrench or wrenches. Some machines require the use of two wrenches, while others use one wrench and a lock button; the lock button holds the arbor stationary while you loosen the arbor nut with the wrench.

Remove the arbor washer and pull the blade off the arbor. Inspect the arbor washers for imperfections (Illus. 604). Flatten them on a sharpening stone if necessary. Place the sharp blade on the arbor. Make sure that the teeth at the bottom of the blade are pointing towards the fence. Also, be certain that the arbor hole in the blade fits snugly over the arbor. Avoid blades with a loose fit.

Replace the arbor washer and the arbor nut (Illus. 605). Tighten the arbor nut securely; if it is not secure, the braking system can loosen it. Replace the guard or the access cover and inspect the movement of the lower guard (Illus. 606). Plug in the saw and make a test cut.

Illus. 604. Always inspect the arbor washers for imperfections. A blade will not orbit in a true plane if the arbor washers are nicked or dented.

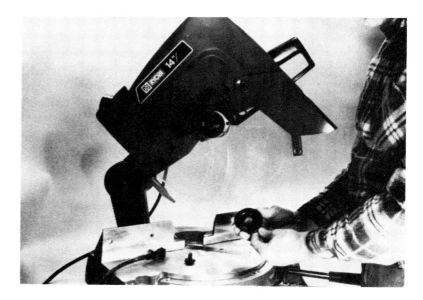

Illus. 605. Replace the arbor washers correctly. They should hold the blade snugly between them.

Illus. 606. Replace the upper guard or close the access cover on the side of the upper guard.

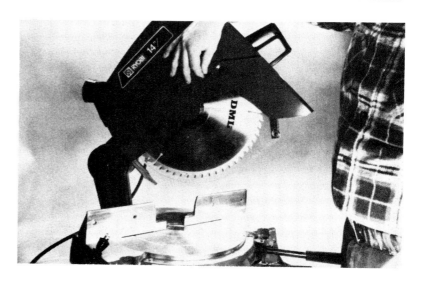

Table Maintenance When wooden tables have been cut several times, they must be replaced (Illus. 607). In most cases, you can do this by removing the four table bolts. Use the old table as a model for a new one. It will help you locate the holes. Select a clean, true piece of stock for your new table. Make sure that it is the proper thickness. Drill the holes carefully so that the fit will be correct (Illus. 608). It is possible to make the table longer than the machine for extra support of the work, but machine storage and transportation is more difficult.

When you are replacing the table, align it with the fence and table. Secure the table snugly and make the crosscut kerf in the table. Disconnect the power and check alignment. The tabletop must be square to the blade and table. In addition, the table center must not crown or sag. By tightening and loosening the four table bolts, you can pull the table into the correct plane.

If the table sags or crowns at the center, most machines have a bolt under the table center to support it; you can turn it in or out to raise or lower the table. Work carefully and check your progress with a straightedge.

Metal tables require little maintenance. For best results, they should be kept free of oxidation. Remove oxidation with auto-body rubbing compound, which will polish the table. Protect the polished surface with paste wax. While working on the table of the mitring machine, check the pivoting mechanism for smooth operation. Add needed lubrication if the pivot mechanism is difficult to turn. Note: If slop is evident in the pivoting mechanism, tighten it to reduce wear and realize the full accuracy of the machine.

Illus. 607. A table that has several cuts should be replaced. Cutoffs get trapped in the cuts and can be thrown from the saw.

Illus. 608. Shim the new table so that it's perpendicular to the saw blade. Select true stock of the same thickness as the original piece.

Alignment To align any mitring machine, begin by making sure that the table is square to the blade and fence. Note: Make all alignment checks and adjustments with the power disconnected. Next, make sure that the blade is square with the fence. If not, you will have to adjust the fence on some machines perpendicular to the blade (Illus. 609), while on others it is adjusted at the handle (Illus. 610). On other machines, the motor-mounting bracket has to be pivoted until the alignment is correct. Make sure that the blade is square after making any adjustment (Illus. 611). Sometimes the movement of the bolts can cause the fence or mounting bracket to "walk" during tightening.

On some mitring machines, you will also have to adjust the slot in the turntable for alignment with the stroke of the blade; you can usually do this by loosening the set screws beneath the table. Now turn the table into position and tighten the setscrews.

Heeling When the blade is not square with the table, the mitring machine is heeling. The blade does

Illus. 609. On some machines, you will have to adjust the fence so that it's perpendicular to the blade.

Illus. 610. On other machines, fine adjustment is done at the handle. Turning the setscrew will change the angle between the blade and fence.

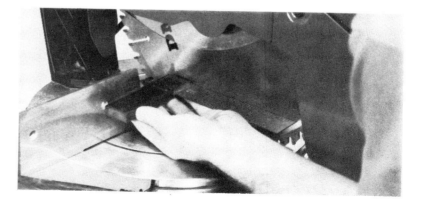

Illus. 611. After making any adjustments, check the angle again to be sure that it is correct.

not come through the workpiece squarely; this can cause rough cuts and tear-out even when the blade is sharp because the teeth are not all in the same plane as the cut being made. To eliminate heel, some systems have provisions to turn the motor and blade unit, while on others you can shim the motor-mounting bracket (Illus. 612). In some cases, only the table can be adjusted. Regardless of which machine you own, you must eliminate heel to achieve the best cuts (Illus. 613).

Check the plunging depth of the blade. If the blade is coming too close to the metal table, adjust the plunging stop (Illus. 614); this will save the table and blade.

Remember, one adjustment often affects another adjustment. Recheck mitring-machine alignment at every step to be sure that the alignment is correct.

Electrical Maintenance While maintaining your chop-stroke mitring machine, check the cord and the brushes. Disconnect the saw before you start. Inspect the cord for cuts or nicks in the insulation. Any cut that breaks through the outer insulation requires cord replacement. Small nicks in the insulation should be patched with electrical tape. Check the plug for wear or oxidation. Remove oxidation with a fine wet/dry abrasive; this will improve electrical contact and reduce the possibility of voltage drop.

Also inspect the brushes for wear. Some mitring machines have access to the brushes from the outside of the case (Illus. 615), while on others you have to remove the motor cap. When the brushes become short, it is time to replace them (Illus. 616). Some brushes have a replacement line on them, while others do not. Usually the brushes should be replaced when they are about ¼ inch long. Replacement brushes will have to be run for a while before the saw will run smoothly and the electric brake will work properly. Always replace brushes as a pair.

Illus. 612. On some machines, the post can be shimmed to eliminate heeling.

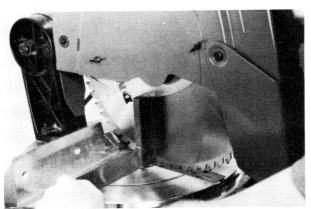
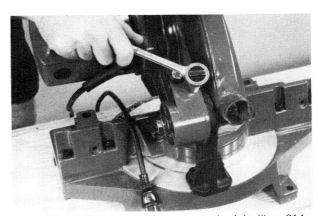

Illus. 613 (above left). Check the setting after making any adjustment. This will ensure quality cuts on the job. Illus. 614 (above right). You can adjust the plunging stop to prevent contact between the blade and the base.

Illus. 615. Brush access is usually on the outside of the saw. Remove the caps with a screwdriver.

Illus. 616. Inspect the brushes for wear. When brushes become shorter than ¼ inch, they should be replaced. Always replace brushes as a pair.

Changing the Belt On motor-driven mitring machines, the cog belt may occasionally require replacement. Disconnect the power before you begin this job. Remove the belt guard and inspect the belt. If the belt shows wear, slide it off the gear and motor shaft. Order and install the appropriate belt. Make sure that it is properly engaged with both the gear and motor shaft before replacing the belt guard.

Purchasing a Chop-Stroke Mitring Machine

When you are purchasing a chop-stroke mitring machine, many factors could influence your decision. The major factors are listed on the planning sheet (Table 3). Use this sheet to help you organize your thoughts. In the following paragraphs I discuss the factors listed on the planning sheet.

Your decision on whether to buy a compound- or simple-mitring machine may be influenced by the type of work you do and its size. The blade on the compound-mitring machine is about 8 inches in diameter. Many of the simple-mitring machines use blades 10–14 inches in diameter. These machines will have a greater stock capacity.

The quality of the blade offered is also important. Some machines offer a carbide-tipped blade as standard equipment; this could mean a dramatic savings.

Generally, electric brakes are preferred to mechanical brakes because they are faster acting and do not wear quickly. The type of table is a personal preference. Some woodworkers like a wooden table because they can nail cleats and other devices to it. The plastic insert in the metal table is preferred by some because it offers splinter-free cutting. The plastic backs up both sides of the cut as the blade goes through.

The type of guard used is also a personal preference. As long as the guard feels right to you, it will probably work. It should be easy to use, protect your hands, and not impair your vision.

The type of bearing is also an important factor. Generally, ball and roller bearings are the best. Sleeve bearings tend to get sloppy with wear.

A final consideration is the accessories that are

available or furnished with the tool. When accessories are available, they tend to make the job easier and more efficient. Retrofitting devices can be time-consuming, and they are not usually as accurate or efficient.

Specialty Mitre Machines

Some motorized mitre boxes have a special shape or perform specific jobs. Some are designed for stationary use (Illus. 617). These machines are very heavy and rigid. They are capable of cutting wood, plastic, or metal.

Another version of this heavy-duty saw is equipped with a small table saw for ripping stock (Illus. 618). This machine could be useful for specialty jobs in the shop. For complete accuracy when cutting mitres, use a double-mitre machine (Illus. 619). Each stroke of the machine cuts two mitres. Pneumatic clamps hold the stock stationary during the cut.

Chop Stroke Mitring Machine Planning Sheet

(Check One)

Compound Mitring _____   Simple Mitring _____

Capacity (Largest Stock Cut)   _____ Tx _____ Wx _____ L

Blade Diameter Needed _____   RPMS _____

Quality of Blade Offered (Check One). Tool Steel _____   Carbide Tipped _____

Brakes (Check One)

    Electrical _____   Mechanical _____

Angle Calibrations

    Moulded into Base _____   Riveted to Base _____

Table (Check One)

    Wood _____   Metal _____   Metal w/Plastic Insert _____

Overall Weight _____

Guard Evaluation (Check One)

                    Poor   Fair   Good

Protects hand
Easy to Use
Provides Good Visibility

Bearing Type (Check One)

    Sleeve _____   Roller _____   Ball _____

Accessories Available (List Price if Available)

    Clamps _____   Table Extensions: _____

    Length Stops _____   Dust Bag _____

Table 3. Use a planning sheet when trying to decide what type of chop-stroke mitring machine to purchase.

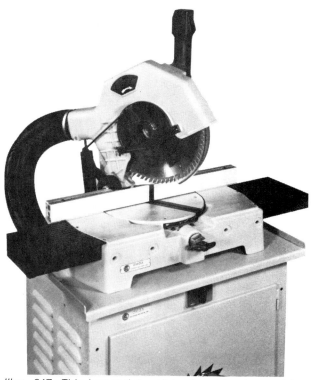

Illus. 617. This heavy-duty mitring machine is used in stationary operations. Note the large dust-collection system on the machine.

Illus. 618. This motorized mitre box is also capable of ripping stock.

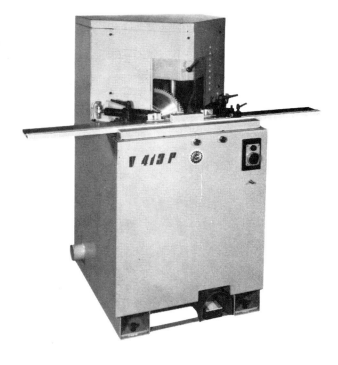

Illus. 619. This double-mitre machine cuts two mitres with each stroke. Pneumatic clamps hold the stock while it is being mitred.

9

Pull-Stroke Mitring Machines

Pull-stroke mitring machines have a motor similar to that of a portable circular saw. This motor travels on a pair of metal rods; it is pulled into the workpiece for cutting. The pull-stroke mitring machine has many of the advantages of the radial-arm saw. It will make many of the cuts the radial-arm saw makes, and has the advantage of being portable.

Pull-stroke mitring machines are very useful for trim work. They can cut boards much wider than a chop-stroke mitring machine. This makes them useful for interior or exterior residential and commercial trim work.

Currently, there are two pull-stroke mitring machines being produced: The Craftsman Compound-Cut Miter Saw and The Delta Sawbuck™ Frame and Trim Saw. The Sawbuck Saw has a much larger capacity than the Compound Cut Saw. The Compound Cut Saw must be mounted on a stand or suitable base. Many of the accessories used with chop-stroke mitring machines can also be used with the Compound Cut Saw.

The Sawbuck is a self-contained saw system. It unfolds to make a table on which large boards can be supported. In addition, it has a table extension that extends from the saw base (Illus. 620 and 621); with this extension, the Sawbuck can support larger boards. When the Sawbuck is folded up, it has a pair of wheels that allow it to be handled like a cart (Illus. 622).

The Sawbuck uses an 8-inch-diameter saw blade, and the Compound-Cut Saw uses a 7½-inch diameter blade. The Sawbuck and the Compound Cut Saws both turn 5,500 rpm's. Both machines seem to have adequate power for the jobs they are intended to perform, and both are equipped with a brake. The Sawbuck has an electric brake (Illus. 623), while the Compound Cut Saw has a manual brake. The brakes on both machines are located above the pull handles, and are usually depressed with the thumb (Illus. 624).

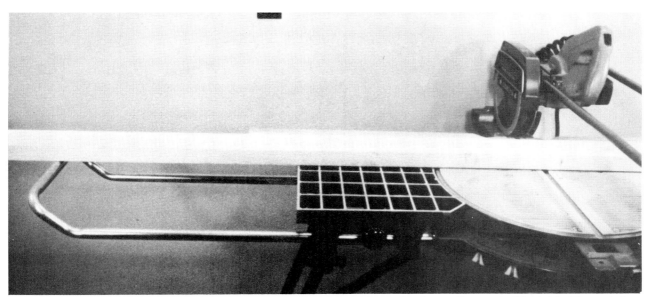

Illus. 620. The extension rod mounted under the saw table supports heavy loads on either end of the table. The rod also doubles as a handle when the Sawbuck is moved.

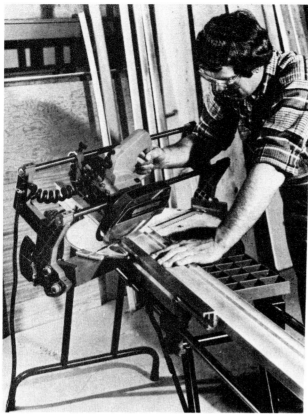

Illus. 621. The rods extending from the Sawbuck support long stock while it is being cut.

Illus. 622. The Sawbuck has a pair of wheels for portability. It can be wheeled to the job with little effort.

Illus. 623. The Sawbuck has an electric brake that you can actuate by depressing the thumb switch.

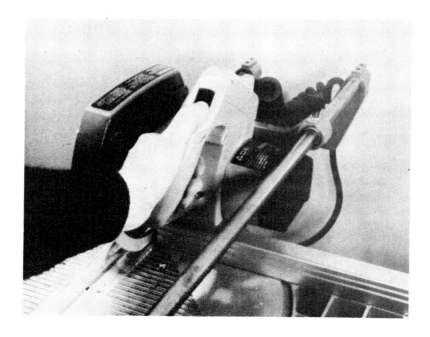

Illus. 624. The Compound Cut Saw has a manual brake. Pressure applied to the thumb button causes the friction needed to stop the blade.

Parts

Both pull-stroke machines have similar parts and elements. Study Illus. 625–630 to identify the parts as they are discussed. The motor and carriage travel on a pair of metal rods. The metal rods are visible on the Sawbuck machine, but are concealed in the swing arm of the Compound Cut Saw.

The upper blade guard on both machines is attached to the motor and carriage. The lower guard on the Sawbuck is made of soft metal; it lifts itself over the work when the carriage is pulled into the work (Illus. 626). The Compound Cut Saw has a clear-plastic lower guard. It, too, lifts over the work as the carriage is pulled into the work (Illus. 627). The upper guard on both machines has a side plate that must be removed to change the blade.

Both carriages can be held stationary with a lock knob. The lock knob or lever is located on the swing arm of the Compound Cut Saw (Illus. 628). On the Sawbuck, the lock knob is located on the right side of the carriage (Illus. 629). It clamps the carriage to the right rod.

The fences and table on the Compound Cut Saw or Sawbuck control the work while it is being cut. The fences on the Compound Cut Saw remain stationary after adjustment. The fences on the Sawbuck must be released when the base and rods are turned or pivoted (Illus. 630). For certain cuts, the fence must be moved to a second position. The table on both machines has a

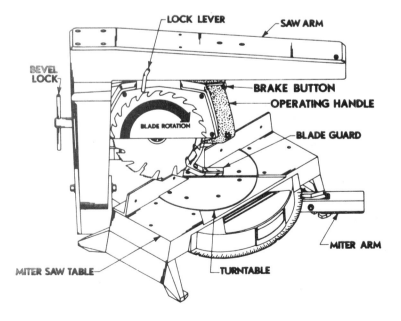

Illus. 625. This drawing will help you familiarize yourself with the parts of the Compound Cut Saw.

Illus. 626. The lower blade guard lifts itself over the workpiece as the cut progresses.

Illus. 627. The lower blade guard rides on the workpiece. The end of this guard is designed to lift over the workpiece.

Illus. 628. You can tighten the lock knob to hold the carriage stationary. Always tighten it when the machine is not being used or being transported.

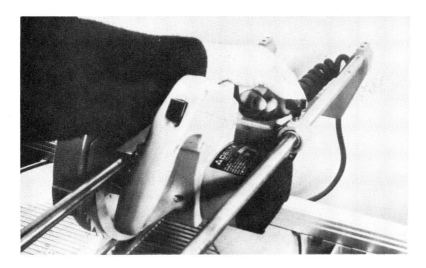

Illus. 629. Thread the lock knob into the right side of the carriage on the Sawbuck. The lock knob holds the carriage stationary for transportation or when the saw is not being used.

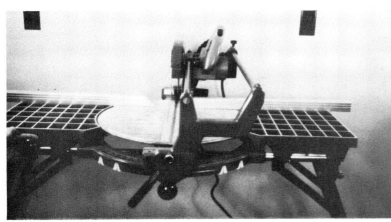

Illus. 630. Pulling out the fence-release knob allows the fences to move when you turn the saw for mitre-cutting.

pivoting portion that turns for angular cuts; this portion supports both sides of the work as the kerf is cut.

To pivot the table, release the knob at the front of the machine. This knob is located at the base of the rod supports on the Sawbuck (Illus. 631). The mitre arm on the Compound Cut Saw has a lock lever above the handle (Illus. 632) and a clamp mechanism beneath the handle (Illus. 633). The clamp locks the turntable at an odd angle. You have to screw the knob on the Sawbuck into the casting to clamp an odd angle (Illus. 634). There are positive index stops on the Sawbuck and Compound Saws. These stops are located at common mitre settings such as 45 degrees.

On the Sawbuck, a blade-pivoting mechanism is located adjacent to the turntable knob. The mechanism is a clamping device (Illus. 635). When you release it, you can tilt the blade. To tilt the blade to the desired angle on the Compound Cut Saw, release the bevel lock on the back of the machine (Illus. 636). When tilting the blade on either machine, make sure that the clamping mechanism is locked securely at the desired angle (Illus. 637). This will make the cut safer and more accurate.

General Safety Guidelines

1. Make sure that the tool is grounded properly. Never work in wet or damp areas.
2. Butt stock against the fence for cutting. Position warped or twisted stock so that it will not pinch the blade as the cut is being made.
3. The blade must be clear of the work when you start the machine. Allow the blade to come to full speed before making a cut.
4. Make all cuts with a pull stroke. Always return the carriage to the rear of the machine when the cut is complete. Lock the carriage at the rear of the saw when it is not in use.
5. Be sure to stop the blade on the Compound Cut Saw by using the brake. A coasting blade can cut you. The electronic brake on the Sawbuck stops the blade automatically after the brake has been pushed.

Illus. 631. You have to release the knob at the front of the machine to turn the saw for mitring. You also have to release the fences before turning the saw carriage.

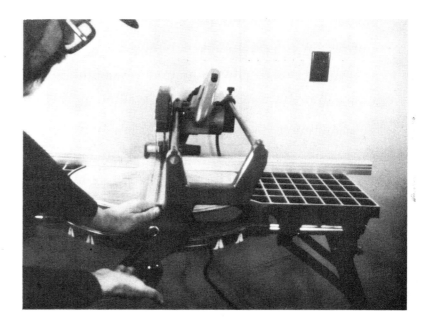

Illus. 632. Depress the lock lever to turn the carriage for a mitre cut.

Illus. 633. The clamping mechanism under the handle locks the turntable at an odd angle.

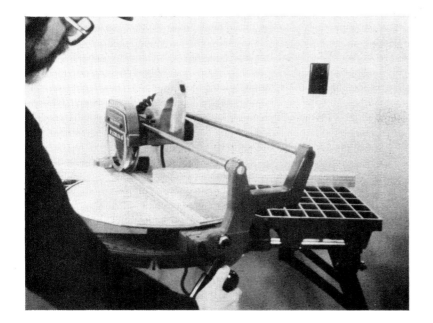

Illus. 634. Tighten the lock knob securely to lock an odd angle. Common angles have a positive stop for quick setting.

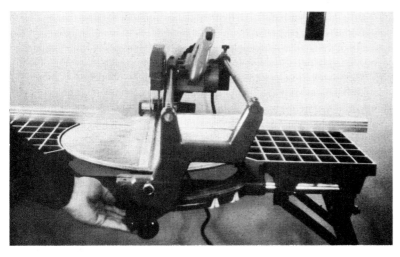

Illus. 635. To pivot the blade, release this clamping mechanism. Pull it straight up to release it.

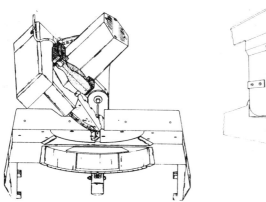

Illus. 636A (above left). To tilt the blade on the Compound Cut Saw, release the clamp knob on the back of the machine. Illus. 636B (above right). The scale at the rear of the machine tells you the angle setting. Lock the knob securely after adjusting the angle.

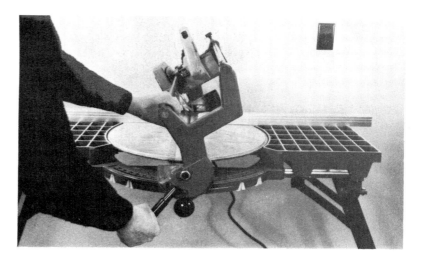

Illus. 637. After the desired angle is set, be sure to lock the clamping mechanism securely. Check it occasionally while working.

6. Keep your hands parallel to and clear of the blade's path. Never cross them.

7. Make all adjustments and change blades with the power disconnected. Make sure that the saw is adjusted according to manufacturer's specifications before you begin using the machine.

8. Make sure that all locks and clamps are engaged and/or clamped securely before making any cuts. Movement of any adjustment during cutting will affect accuracy and could cause an accident.

9. Support long pieces of stock. Use sawhorses or other support devices to hold stock in a true plane. Failure to support long pieces could cause them to lift into the blade during the cut. Avoid twisted or warped stock. Use a portable circular saw to cut twisted stock into smaller pieces.

10. Keep a 6-inch margin of safety from the blade and never hand-hold stock less than 12 inches long.

11. Always clamp round stock to keep it from rolling into the blade during a cut.

12. Check the arbor periodically while working with the Sawbuck or Compound Cut Saw. The sudden stops made by the braking system can cause the arbor bolt to work loose.

13. Analyze all jobs before you begin working. Ask yourself "What will happen when I" . . . or, "Is there a safer way to do the job?" If you feel that the job is unsafe, do not begin. Ask an experienced woodworker if there is a safer way. Remember, many people who have woodworking accidents have had a premonition of danger before starting the job. Trust your intuition and avoid situations that make you feel unsafe.

Common Operations

Pull-stroke mitring machines have many common operations; they include crosscutting, mitring, and compound-mitring. These and related topics are discussed in the following paragraphs:

Crosscutting Begin crosscutting by butting the stock firmly against the fence. Position the layout line with the blade on the waste side (Illus. 638 and 639). Use a tooth that points towards the layout line to position the work. Hold stock against the fence with your left hand, while using your right hand to turn on the saw and pull it through the work (Illus. 640). Keep your right arm slightly stiff as you pull the carriage into the work (Illus. 641). In some cases the blade will grab the work and try to climb into it. A stiff arm can resist this climbing force. The thicker the stock, the greater the tendency of the blade to climb the work (Illus. 642 and 643).

As soon as you cut the work, shut off the saw. On the Compound Cut Saw, apply the brake (Illus. 644 and 645). You have to press the braking button on the Sawbuck to start the electric brake. As the blade stops, return the carriage to the rear of the saw. Lock the carriage in position if cutting is complete (Illus. 646 and 647).

When you are crosscutting round stock, there is a tendency for the blade to climb into it in a gear-like manner. It is important that you securely clamp the round stock (Illus. 648 and 649) to avoid this gear-like effect. Failure to do so could jam the machine and throw it out of adjustment. It could also cause the tips on a carbide blade to break or shatter. It is very expensive to replace carbide tips on a blade (Illus. 650).

It is best to avoid cutting twisted or warped stock on either the Compound Cut Saw or Sawbuck. If they must be cut, it is important that you butt the stock securely to the fence near the cut (Illus. 651). This keeps the stock from pinching the blade as the cut is being made. If possible, cut large twisted pieces into

Illus. 638 (above left). Position the workpiece so that the layout line is aligned with the blade and the stock is butted against the fence. Illus. 639 (above right). Position the saw blade on the waste side of the layout line. Make sure that the stock is butted against the fence.

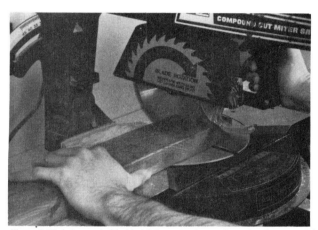 

Illus. 640 (above left). Hold the stock against the fence with your left hand and pull the carriage with your right hand. Illus. 641 (above right). Keep your right arm stiff to control saw climbing.

Illus. 642 (above left). Thicker stock is more likely to cause the saw to climb. Illus. 643 (above right). The cut is complete when the center of the saw blade lines up with the end of the workpiece.

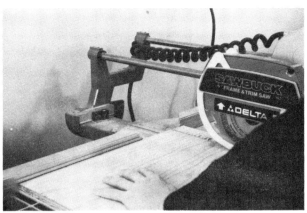 

Illus. 644 (above left). When the cut is complete, release the trigger. Illus. 645 (above right). Then apply the brake to stop the blade.

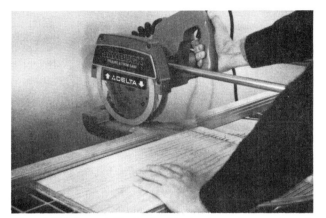 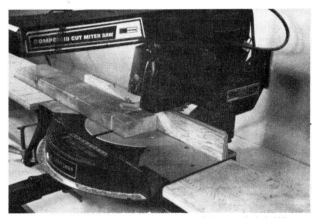

Illus. 646 (above left). As the blade slows down, return the carriage to the rear of the saw. Illus. 647 (above right). When cutting has been completed, lock the carriage in position with the clamping mechanism.

Illus. 648. For best results, clamp round stock when crosscutting.

Illus. 649. The clamp keeps the round stock from acting like a gear and rolling into the blade. This can cause blade damage.

Illus. 650. Clamping keeps the round stock from damaging the carbide tips on the blade. These tips are brittle and can be broken easily if the stock rolls into the blade.

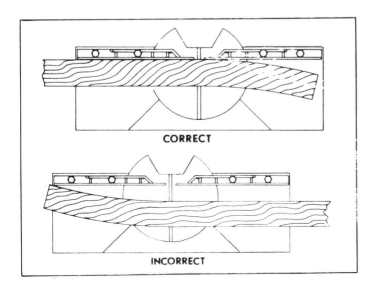

Illus. 651. Warped or twisted stock should be butted securely to the fence where the cut is to be made. This keeps the work from pinching the blade.

smaller parts with a portable circular saw. This will minimize the amount of twist or warp.

When crosscutting several parts to length, you can use a stop to position the work. The Sawbuck has an adjustable stop that can be clamped to the fence (Illus. 652). You can clamp a stop to the fence of the Compound Cut Saw or any extension table used with it (Illus. 653). Measure carefully when setting a stop in position (Illus. 654). Check the setting on scrap before you begin cutting stock. Do not ram the stock against the stop while you are cutting (Illus. 655 and 656). This could cause the stop to creep and change the distance between it and the blade.

When crosscutting long pieces, keep the work supported with an auxiliary table or other support (Illus. 621 and 657). It is unsafe to attempt to hold heavy pieces against the table without support. If the piece gets away from you, it could force your hand into the blade. Control of the stock is essential to safe operation of any mitring machine.

A table extension can be fastened to the table of the Compound Cut Saw (Illus. 658). This reduces the fence height, but improves control of the stock. Stock and stops can be clamped to the table extension (Illus. 659).

The table extension is also useful for cutting thin stock; it lifts it above the joint between the table and fence where it could be pinched or creep under the fence (Illus. 660). An auxiliary table can also be fabricated for the Sawbuck when you are cutting thin stock.

Illus. 652. An adjustable stop is furnished with the Sawbuck. It clamps to the fence for repetitive cutting of stock of any dimension.

Illus. 653. You can clamp a stop to the extension table to control stock length.

Illus. 654. Measure carefully from the stop to the blade for accurate adjustment. Clamp the stock securely in position.

Illus. 655. Stock is held against the fence and stop while the cut is being made. Avoid ramming the stop when moving the work. This could affect the accuracy of the setup.

Illus. 656. When cutting multiple parts, stack them clear of the cutting area. A neat work area makes the work safer.

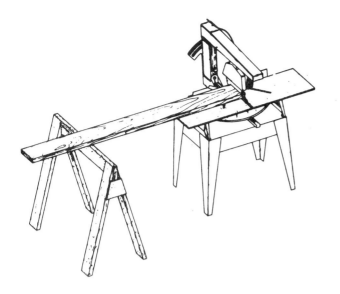

Illus. 657. Support long pieces with a sawhorse or other work support. Control of the work is essential for safe cutting.

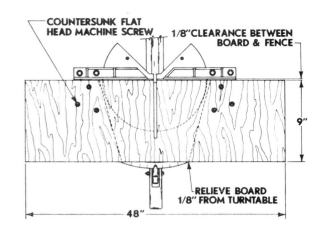

Illus. 658. An extension table can be fabricated for the Compound Cut Saw. Screw it to the metal saw table.

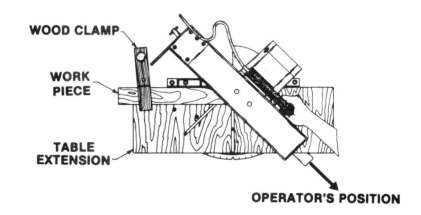

Illus. 659. You can clamp the work or stop to the auxiliary table.

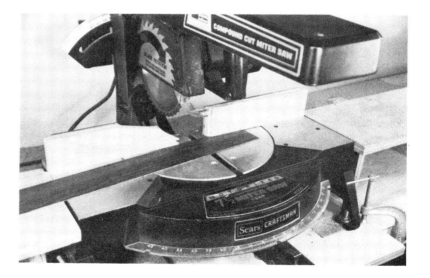

Illus. 660. Thin stock can sometimes slip between the table and fence. An auxiliary table will eliminate this problem.

Cutting Face Mitres Mitres across the face of the work are made with the blade perpendicular to the table. Turn the track or mitre arm to the desired angle (Illus. 661 and 662). There are positive stops at 45 degrees that make the setup of conventional mitres easier. Lock and/or clamp the mitre setting securely before doing any cutting (Illus. 663 and 664). When working with short pieces, clamp them to the fence or table for cutting. Butt stock against the fence and hold it securely on the table (Illus. 665).

Position your hand at least ten inches from the blade. Remember, if the blade were to grab the work, it could pull the work and your hand towards it while it turns. If you place the work to the left of the blade when the arm is turned to the right, this climbing condition is not likely to occur (Illus. 666 and 667). The same is true when you turn the arm to the left, and place the work to the right of the blade. However, some jobs do not lend themselves to the above procedure.

Work carefully with your hands well away from the blade. Remember that the scrap or cutoff can also get caught in the blade. These pieces are sometimes thrown by the blade.

Make the mitre cut the same way as you would make a crosscut (Illus. 668 and 669). Observe the same safety practices. This includes supporting long or heavy pieces of stock.

As the cut is completed, turn off the saw. Let the blade come to a complete stop and return the carriage to the column (Illus. 670 and 671). Clamp it to the column when cutting is complete (Illus. 672).

Work carefully and concentrate on the job. Keep your hands clear of the blade's path. Never cross them when using a pull-stroke mitring machine.

Illus. 661. When face-mitring, turn the mitre arm to the desired angle.

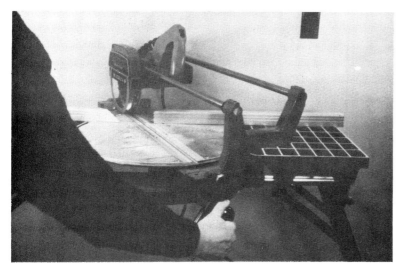

Illus. 662. Release the fence and turn the track arm to the desired angle. Clamp the setting securely.

Illus. 663. Clamp the mitre arm after setting the desired angle. The lock is under the arm handle.

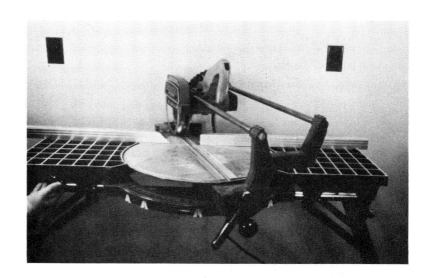

Illus. 664. Lock the fences in position after setting the track arm to the desired angle.

Illus. 665. Before mitring the work, make sure that it is butted firmly against the fence. The stock should not rock on the table.

Illus. 666. Hold the work with your left hand and pull the carriage with your right hand.

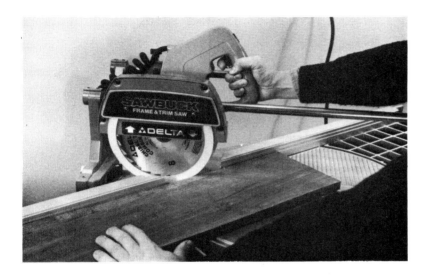

Illus. 667. Keep your right arm stiff to avoid climbing.

Illus. 668 (above left). Make the mitre cut the same way you would make a crosscut. Observe the same procedures. Illus. 669 (above right). When making this mitre cut, observe the same safety practices you would for crosscutting.

Illus. 670. When you have completed the mitre cut, let the blade come to a complete stop.

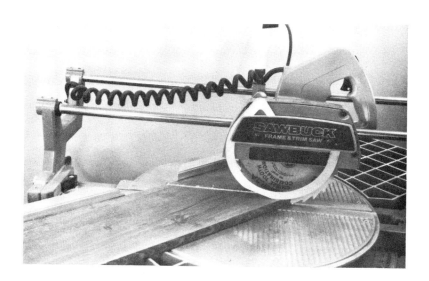

Illus. 671. Apply the brake.

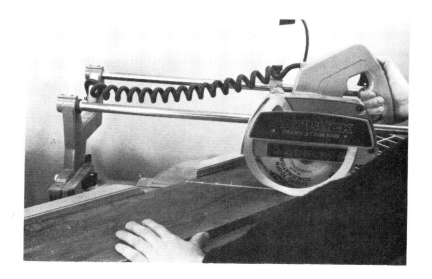

Illus. 672. With the brake applied, return the carriage to the rear of the machine. Lock it in position if cutting is complete.

Cutting End Mitres Mitres across the end of the work can be cut in one of two ways. On narrow boards, cut an end mitre with the blade perpendicular to the table and the track or mitre arm turned to 45 degrees (or other desired angle). Lock and clamp all settings securely. Turn the stock on edge (Illus. 673). The procedure is the same as that used for cutting a face mitre (Illus. 674–678).

Mitre wide pieces of stock by tilting the blade to 45 degrees or another desired angle and positioning the track or mitre arm for crosscutting (perpendicular to the fence). Lay the face of the stock flat on the table for cutting. Make the cut the same way as a crosscut (Illus. 679–684). Remember to clamp short stock securely and to support long or heavy stock when end-mitring.

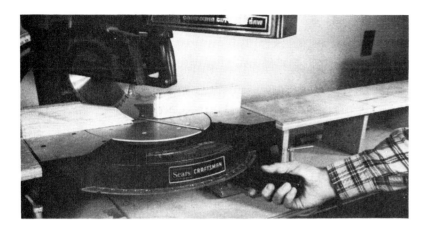

Illus. 673. Make the setting for the end mitre and clamp it securely.

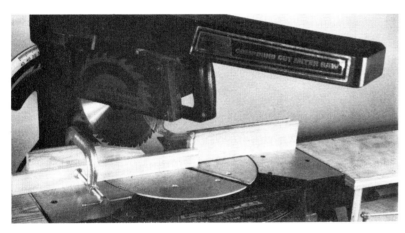

Illus. 674. To end-mitre small pieces, clamp them to the fence. Make sure that the layout line is aligned with the blade.

Illus. 675. Grasp the carriage firmly and turn on the saw. Keep your arm slightly stiff.

Illus. 676. Pull the blade into the work at a steady speed. Keep your hands clear of the blade's path.

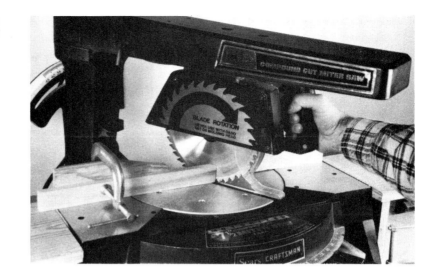

Illus. 677. When the center of the blade comes through the work, turn off the saw and return it to the rear of the saw.

Illus. 678. For some end mitres, use an auxiliary table to elevate the stock. This allows the machine to cut a wider end mitre.

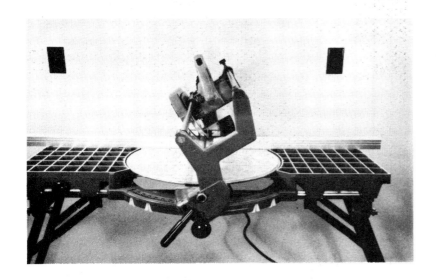

Illus. 679. Begin an end mitre on the Sawbuck by turning the blade to the desired angle. Lock the setting securely.

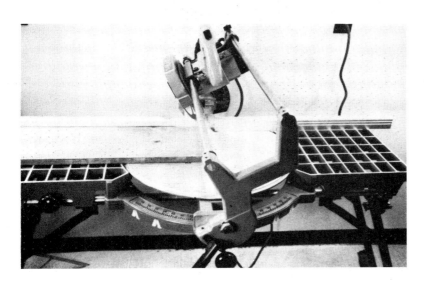

Illus. 680. Line up the cutting line with the saw blade. Make sure that the work is butted securely to the fence.

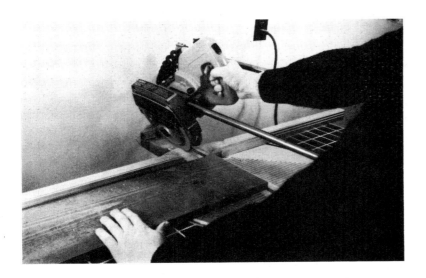

Illus. 681. Turn on the saw and pull it into the workpiece. Keep your left hand well away from the blade's path.

Illus. 682. Keep your right hand slightly stiff as you begin the cut. Pull the carriage at a steady speed.

Illus. 683. When the cut is complete, turn on the brake and return the saw to the back of the machine.

Illus. 684. Lock the carriage in position at the rear of the saw.

Cutting Compound Mitres Mitres with two angles are known as compound mitres or hopper cuts. When four parts are assembled in a funnel or hopper shape, compound mitres are cut on both edges of all four parts. Compound mitres are also cut on the ends of several types of rafters and on crown moulding used between finished ceilings and walls.

When making hoppers, use the chart of mitre and bevel settings shown in Illus. 685 to set up the saw for the cut. Lock both settings securely before making any cuts (Illus. 686–689).

When making hoppers, remember that mitre and bevel angles are very important for close fits. Hopper angles can be correct and the fit will still be poor. If this occurs, all four parts may not be equal in length.

Use scrap stock for setup purposes. Measure all

PITCH OF SIDE	NUMBER OF SIDES						
	4	5	6	7	8	9	10
0°	M-45.00° B- 0.0°	M-36.00° B- 0.0°	M-30.00° B-00.0°	M-25.71° B- 0.0°	M-22.50° B- 0.0°	M-20.00° B- 0.0°	M-18.00° B- 0.0°
5°	M-44.89° B- 3.53°	M-35.90° B- 2.94°	M-29.91° B- 2.50°	M-25.63° B- 2.17°	M-22.42° B- 1.91°	M-19.93° B- 1.71°	M-17.94° B- 1.54°
10°	M-44.56° B- 7.05°	M-35.58° B- 5.86°	M-29.62° B- 4.98°	M-25.37° B- 4.32°	M-22.19° B- 3.81°	M-19.72° B- 3.40°	M-17.74° B- 3.08°
15°	M-44.01° B-10.55°	M-35.06° B- 8.75°	M-29.15° B- 7.44°	M-24.95° B- 6.45°	M-21.81° B- 5.68°	M-19.37° B- 5.08°	M-17.42° B- 4.59°
20°	M-43.22° B-14.00°	M-34.32° B-11.60°	M-28.48° B- 9.85°	M-24.35° B- 8.53°	M-21.27° B- 7.52°	M-18.88° B- 6.72°	M-16.98° B- 6.07°
25°	M-42.19° B-17.39°	M-33.36° B-14.38°	M-27.62° B-12.20°	M-23.58° B-10.57°	M-20.58° B- 9.31°	M-18.26° B- 8.31°	M-16.41° B- 7.50°
30°	M-40.89° B-20.70°	M-32.18° B-17.09°	M-26.57° B-14.48°	M-22.64° B-12.53°	M-19.73° B-11.03°	M-17.50° B- 9.85°	M-15.72° B- 8.89°
35°	M-39.32° B-23.93°	M-30.76° B-19.70°	M-25.31° B-16.67°	M-21.53° B-14.41°	M-18.74° B-12.68°	M-16.60° B-11.31°	M-14.90° B-10.21°
40°	M-37.45° B-27.03°	M-29.10° B-22.20°	M-23.86° B-18.75°	M-20.25° B-16.19°	M-17.60° B-14.24°	M-15.58° B-12.70°	M-13.98° B-11.46°
45°	M-35.26° B-30.00°	M-27.19° B-24.56°	M-22.21° B-20.70°	M-18.80° B-17.87°	M-16.32° B-15.70°	M-14.43° B-14.00°	M-12.94° B-12.62°
50°	M-32.73° B-32.80°	M-25.03° B-26.76°	M-20.36° B-22.52°	M-17.20° B-19.41°	M-14.91° B-17.05°	M-13.17° B-15.19°	M-11.80° B-13.69°
55°	M-29.84° B-35.40°	M-22.62° B-28.78°	M-18.32° B-24.18°	M-15.44° B-20.82°	M-13.36° B-18.27°	M-11.79° B-16.27°	M-10.56° B-14.66°
60°	M-26.57° B-37.76°	M-19.96° B-30.60°	M-16.10° B-25.66°	M-13.54° B-22.07°	M-11.70° B-19.35°	M-10.31° B-17.23°	M- 9.23° B-15.52°
65°	M-22.91° B-39.86°	M-17.07° B-32.19°	M-13.71° B-26.95°	M-11.50° B-23.16°	M- 9.93° B-20.29°	M- 8.74° B-18.06°	M- 7.82° B-16.26°
70°	M-18.88° B-41.64°	M-13.95° B-33.53°	M-11.17° B-28.02°	M- 9.35° B-24.06°	M- 8.06° B-21.08°	M- 7.10° B-18.75°	M- 6.34° B-16.88°
75°	M-14.51° B-43.08°	M-10.65° B-34.59°	M- 8.50° B-28.88°	M- 7.10° B-24.78°	M- 6.12° B-21.69°	M- 5.38° B-19.29°	M- 4.81° B-17.37°
80°	M- 9.85° B-44.14°	M- 7.19° B-35.37°	M- 5.73° B-29.50°	M- 4.78° B-25.30°	M- 4.11° B-22.14°	M- 3.62° B-19.68°	M- 3.23° B-17.72°
85°	M- 4.98° B-44.78°	M- 3.62° B-35.84°	M- 2.88° B-29.87°	M- 2.40° B-25.61°	M- 2.07° B-22.41°	M- 1.82° B-19.92°	M- 1.62° B-17.93°
90°	M- 0.00° B-45.00°	M- 0.00° B-36.00°	M- 0.00° B-30.00°	M- 0.00° B-25.71°	M- 0.00° B-22.50°	M- 0.00° B-20.00°	M- 0.00° B-18.00°

Each B (Bevel) and M (Mitre) Setting Is Given to the Closest 0.005°.

COMPOUND-ANGLE SETTINGS FOR POPULAR STRUCTURES

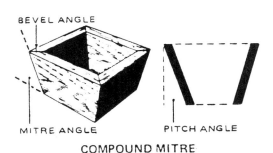

COMPOUND MITRE

Illus. 685. Use this chart and illustration to determine the correct settings for compound-mitring.

Illus. 686. Butt the stock against the fence with the layout line aligned with the blade.

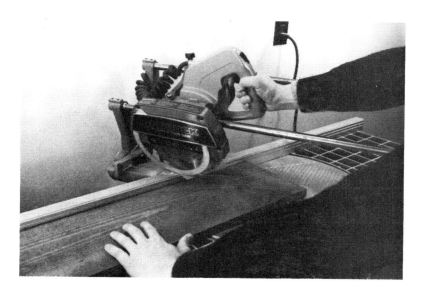

Illus. 687. Cut a compound mitre the same way you would a flat mitre. Allow the blade to come up to full speed before you begin cutting.

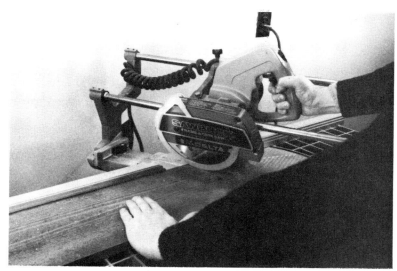

Illus. 688. Turn on the brake when the compound mitre is completed.

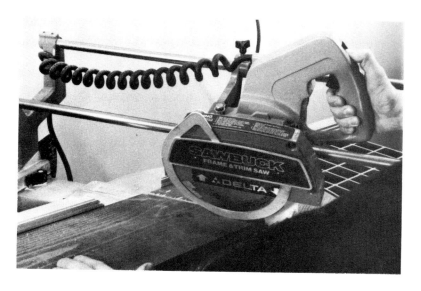

Illus. 689. Lock the carriage at the rear of the machine when the cut is completed.

parts carefully. If parts are small, clamp them to the fence or table for cutting. Keep your hands clear of the blade's path.

Crown mouldings require special settings for accurate work. On the Compound Cut Saw, turn the bevel post to 33.85 degrees and the mitre arm to 31.62 degrees right or left, depending on what cut is being made (Illus. 690–694). On the Sawbuck, set the blade bevel to 33⅞ degrees and turn the arm to 31⅝ degrees right or left, depending on what cut is being made (Illus. 695 and 696).

The settings furnished for crown mouldings are only indicators. The finished walls and ceilings may not intersect at a 90-degree angle. When this happens, make cuts by trial and error. For some inside corners, use a cope joint. The fit is usually tighter, and a corner intersection over or under 90 degrees is not as difficult to work with. Use the chart that appears on page 243 as a guide to setting up the Sawbuck or Compound Cut Saw to make crown moulding cuts.

Rafters are sometimes cut as compound mitres. Use a framing square to lay out the angles on the work.

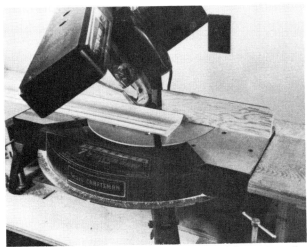

Illus. 690 (above left). To cut crown mouldings, turn the bevel post to 33.85 degrees and set the mitre arm at 31.62 degrees. These angles are not always perfect, but this is a good place to start. Illus. 691 (above right). You can cut an inside-corner left mitre with this setup. The scrap on the right side of the blade is actually an outside-corner right mitre.

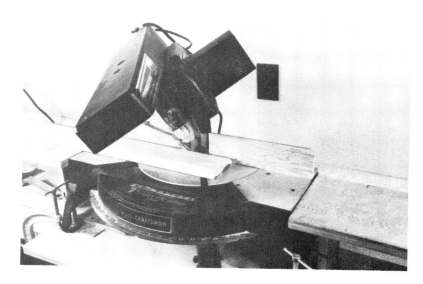

Illus. 692. The setup here is for an outside-corner left mitre.

Illus. 693. Keep stock firmly against the fence when cutting a compound mitre. Stock may tend to creep up as cutting begins.

Illus. 694. A sharp blade is essential for quality mitres. Remember, if floors or walls do not intersect at 90-degree angles, mitre adjustments must be made.

Illus. 695. An outside-corner right mitre is being cut with this setup. Work slowly and carefully when working with cove moulding. It is expensive; waste should be kept to a minimum.

Illus. 696. Compound mitres can be cut accurately and cleanly with all pull-stroke mitring machines.

Inside Corner

Right Mitre—Turn arm to left, good side up — ceiling edge away from fence, stock to left of blade

Left Mitre—Turn arm to right, good side up — ceiling edge towards fence, stock to left of blade

Outside Corner

Right Mitre—Turn arm to right, good side up — ceiling edge against fence, stock to right of blade

Left Mitre—Turn arm to left, good side down — ceiling edge away from fence, stock to right of blade

Scarf Joint—Cut one inside left mitre
Cut one outside right mitre

Table 4. Use this chart to set up the Sawbuck or Compound Cut Saw to make crown moulding cuts.

Copy the angles with a sliding T bevel. Use the sliding T bevel to set the arm and blade bevel. Clamp and lock the settings securely. Support the work with some type of auxiliary table or other device. Make sure that the work is butted to the fence (Illus. 697 and 698). Turn on the saw and make the cut. Hold the carriage securely as you pull it into the work (Illus. 699). Since rafter stock is quite thick, there is a chance that the blade could climb the work. Be sure to keep your hands clear of the blade.

After the cut is complete, turn off the saw. Apply the brake if you are using the Compound Cut Saw (Illus. 700). Return the carriage to the back of the table after the blade has come to a complete stop (Illus. 701).

The opposite ends of some hip or valley (compound-cut) rafters do not need a compound-mitre cut. This is also true of a common rafter. These rafters need a simple mitre (Illus. 702). The angle of the mitre is determined by the rise and run of the rafter. Use a framing square to lay it out (Illus. 703) or consult a table of rafter angles. Complete books have been written on rafter framing; consult one of these books if you are cutting rafters on your pull-stroke mitring machine.

Maintenance of Pull-Stroke Mitring Machines

Changing the Blade When the blade becomes dull, replace it. Begin by disconnecting the machine from its power source. Both pull-stroke mitring machines have an access plate on the upper guard (Illus. 704). It must be removed when you change the blade (Illus.

 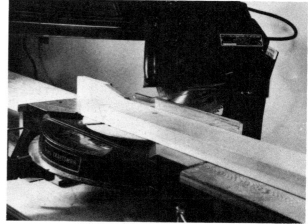

Illus. 697 (above left). Some rafters require a compound-mitre cut. Adjust the layout line to the blade's path. Illus. 698 (above right). Rafter stock or heavier parts must be supported fully while the cut is being made.

Illus. 699. Keep your arm slightly stiff as you pull the blade into the work. There is a tendency for the blade to climb thicker stock.

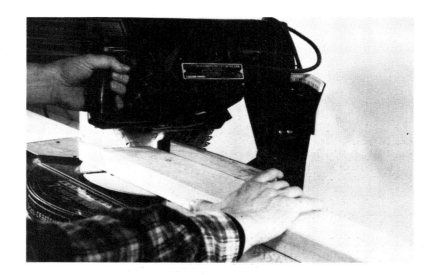

Illus. 700. Apply the brake and return the carriage to the rear of the machine when the cut is completed.

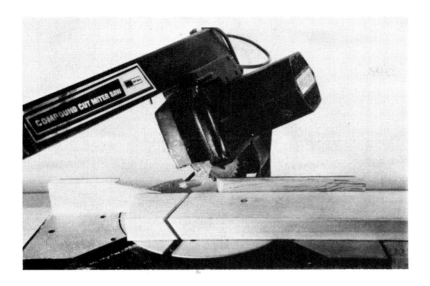

Illus. 701. With careful layout, rafter cuts will be very accurate. Make test cuts in scrap stock when you are unsure of the setup.

Illus. 702. Some rafter cuts are nothing more than a flat mitre cut.

Illus. 703. You can use a framing square to lay out rafter cuts. Careful layout ensures accurate results.

Illus. 704. To change the blade on either pull-stroke mitring machine, you have to remove the upper guard.

Illus. 705. Use a screwdriver to remove the side plate on the upper guard.

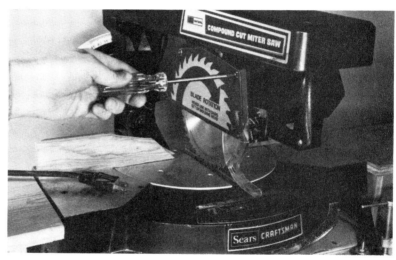

705). Use two wrenches to loosen the arbor bolt on the Sawbuck (Illus. 706), one on either side of the blade. On the Compound Cut Saw, use a block of wood to hold the blade and arbor stationary while loosening the arbor bolt with the wrench (Illus. 707). Both machines have left-hand threads, so turn the wrench clockwise.

Remove the arbor nut and the outer arbor washer. The blade can now be removed. Inspect the inner and outer arbor washers for dents or nicks (Illus. 708). Remove any nicks or dents by rubbing the flat faces on a sharpening stone. Replace the inner arbor washer and mount the new blade (Illus. 709). Replace the outer arbor washer and arbor bolt. Tighten the arbor bolt securely (Illus. 710). Turn the wrench counterclockwise. If the arbor nut is not tightened securely, it could come loose when the brake is applied. Check the arbor bolt periodically for tightness. Replace the access plate and tighten the screws (Illus. 711).

Machine Adjustments For accurate work, you must make three common adjustments on the pull-stroke mitring machine. All of these adjustments should be made with the power disconnected. First, the blade must be square to the table. To check this on either machine, remove the access plate from the upper guard (Illus. 712). Use a square to check the angle between the blade and table (Illus. 713). On the Compound Cut Saw, loosen the lock screw on the column to adjust the angle between the table and the blade. Once the angle is set, tighten the lock screw and adjust the pointer on the angle scale to 0 degrees.

On the Sawbuck, use an Allen wrench to turn the setscrew under the table lock knob. Turn the screw

Illus. 706. You have to use two wrenches to remove the blade from the Sawbuck. The bolt has left-hand threads.

Illus. 707. You only need one wrench for the Compound Cut Saw. Block the blade with wood scrap while loosening the bolt.

Illus. 708. When changing saw blades, check the arbor washers for dents or nicks. For best cutting, remove any dents or nicks.

Illus. 709. Once the blade is mounted, attach the outer arbor washer and bolt.

Illus. 710. Tighten the bolt securely. The braking action has a tendency to loosen the arbor bolt.

Illus. 711. Replace the upper guard after you have mounted the blade. Tighten the screws snugly.

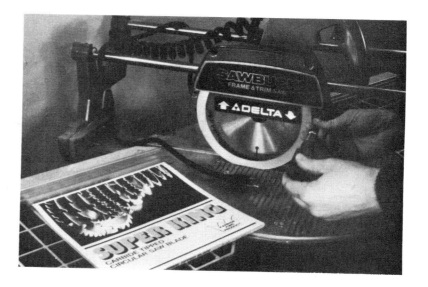

Illus. 712. Remove the side of the upper guard to check the angle between the blade and the table.

Illus. 713. Use a square to determine whether the blade is perpendicular to the table.

until the blade is perpendicular to the table. Move the carriage to the back of the table and check the blade again. Turn a similar setscrew at the other end of the rods to adjust the blade angle at the rear of the table; then adjust the pointer at the opposite end of the rods to 0 degrees. Next, turn the carriage to 45 degrees and check the blade again (Illus. 714). If the angle is incorrect, turn the setscrew above the clamp mechanism (Illus. 715) until the angle is correct. Then move the carriage to the rear of the machine and check the blade again. Turn a similar screw at the other end of the rods to adjust the angle at the rear of the machine.

The second adjustment consists of squaring the blade with the fence. On the Compound Cut Saw, you will do this by loosening the fence and squaring it with the blade (Illus. 716). It is also important that both fences be in a true plane (Illus. 717). Check their alignment with a framing square. Readjust them if necessary. If the angle between the fence and blade is only off a few degrees, you can make a fine adjustment at the mitre detent handle (Illus. 718) by loosening the clamp and then loosening the hex nut. Turn the slotted portion of the screw until the angle is correct, and then tighten the hex nut. Make this adjustment with the locking clamp in the off position.

On the Sawbuck, adjust the angle by loosening the five hex-head bolts under the saw base. Then shift the index unit until the blade travels perpendicular to the fence. Tighten the bolts securely once the angle is correct.

The third adjustment is to eliminate any heeling condition. On the Compound Cut Saw, check for heeling with a combination square (Illus. 719). Butt

Illus. 714. After the blade has been squared to the table, check the 45-degree mitre angle.

Illus. 715. Adjust the Sawbuck by turning the setscrew above the clamp mechanism. Adjust it both at the front and back of the rails.

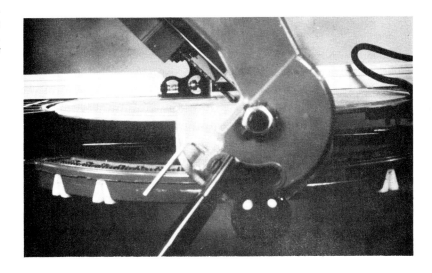

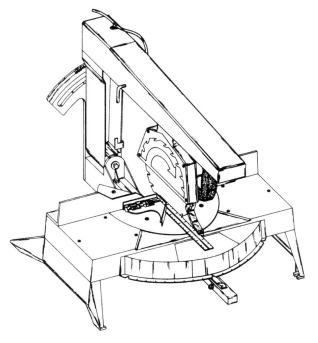

Illus. 716 (left). Square the blade to the fence after you have squared it to the table. Move the fence to make this adjustment. Illus. 717 (above). Check the fences to be sure that they form a straight line. Auxiliary faces can be used to align the two fences.

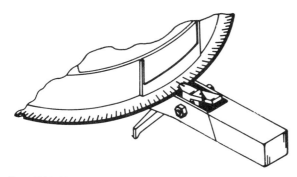

Illus. 718. You can make the fine adjustment of the angle between the blade and fence at the mitre detent.

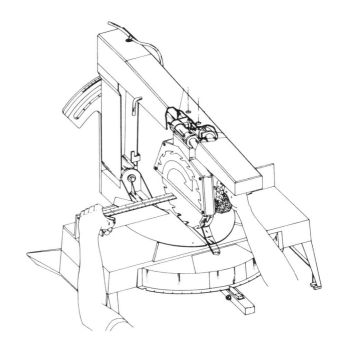

Illus. 719 (right). Check for heeling with a combination square. As the blade passes the square, it should remain a uniform distance from the square. A heeling condition exists if the distance is not uniform.

the blade of the square against the fence, with the end touching the front of the blade. Then slowly push the carriage towards the column. If the blade of the square moves or if a space is observed between the blade and the blade of the square, then a heeling condition exists. The blade is not parallel to its cutting stroke.

To eliminate heel, you have to adjust the carriage. Pull the carriage back until the cap screws are visible through the holes in the overarm. Loosen the cap screws slightly, and turn the carriage until the blade is parallel to its cutting stroke. Tighten the cap screws securely and check the alignment again. Tightening the cap screws can sometimes alter the alignment slightly.

On the Sawbuck, also use a combination square to check for heeling (Illus. 720). Hold the blade of the square against the fence, and butt the head of the combination square against the blade. Any space between the blade and the head of the square means that a heeling condition exists.

To eliminate heeling, loosen the two hex bolts closest to the left rod. Then turn the carriage and readjust it to eliminate the heel. Then tighten the bolts securely. Recheck the setting after tightening the bolts to be sure that the setting remains correct.

Electrical Maintenance Electrical considerations include the cord, plug, and brushes. Always inspect the cord for damage. Any nicks in the insulation should be repaired with electrical tape. Any cut through the outer insulation suggests a need for cord replacement. Also inspect the plug periodically. The plug should not be separating from the cord, and the metal prongs should be free of oxidation.

To inspect the brushes on either saw, begin by disconnecting the power. On the Compound Cut Saw, the brush caps are on the motor housing. Remove the brush caps with a straight-blade screwdriver (Illus. 721). Pull the brushes out and inspect. If they are less than ¼ inch long, or appear burned, they should be replaced. Otherwise, they are still functional (Illus. 722).

Illus. 720. You can also check for heeling on the Sawbuck with a combination square.

Illus. 721. Remove the brush cap on the Compound Cut Saw with a straight-blade screwdriver. Be sure to disconnect the power before removing the brush cap.

Illus. 722. Inspect the brushes carefully. Replace them if they are burned or are less than ¼ inch long.

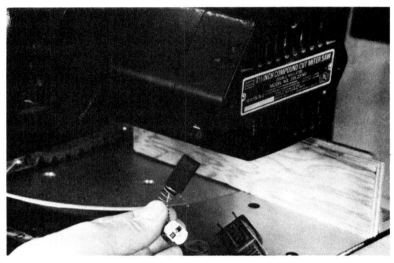

On the Sawbuck, remove the end motor housing and the brush holders to inspect the brushes (Illus. 723). Brushes less than ¼ inch long or with a burned shunt wire should be replaced.

When you are replacing brushes on either machine, replace them as a pair. After installation, turn the saw on and allow it to run for 2–3 minutes; this will allow the brushes to "break in" and fit the armature correctly.

Lubrication The Compound Cut Saw requires no lubrication, but experience suggests that the two rods in the overarm must be lubricated occasionally so that the carriage can travel freely. Paste wax works well as a lubricant on the Compound Cut Saw. It does not attract sawdust. The carriage on the Sawbuck has felt wipers that ride on the metal rods; these wipers must be lubricated occasionally to keep the carriage running smoothly. For best results, keep the rods clean and free of rust. Blow dust off either machine after use. This will keep moisture and rust off the machine.

Accessory Tables

Accessory tables make the pull-stroke mitring machine more useful. They support the stock and increase the accuracy of the cuts. Some accessory tables are sold as commercial products, while others are shop-made. Some commercial components are also sold to help the woodworker make a table using specialty couplers and folding legs.

Some shop-made tables are designed for shop use (Illus. 724). These plywood tables hold stock in a true plane for cutting. The pull-stroke mitre saw has replaced the radial-arm saw for most shop operations (Illus. 725).

Some tables make work in the field easier. The

table shown in Illus. 726. has a sliding stop that makes setup of repetitive cuts much easier (Illus. 727). This table was shop-made with commercial hardware (Illus. 728); those parts were mounted to the shop-built table (Illus. 729 and 730).

If you make accessory tables for your saw, remember that they must be sturdy, easy to assemble, and easy to transport if they are used in the field. Without these features, an accessory table has little value.

Illus. 723. You have to remove the motor housing on the Sawbuck to inspect the brushes. Disconnect the power before doing this.

Illus. 724 (above left). This shop-made table is made of plywood. It has been made to be used in the shop. Illus. 725 (above right). This Compound Cut Saw was set up as a down-sized radial arm saw. The plywood tables support the work at the correct height.

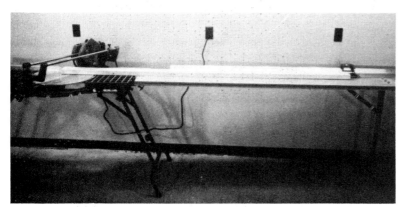

Illus. 726. This table makes work in the field easier. It is a shop-made table that uses commercial components.

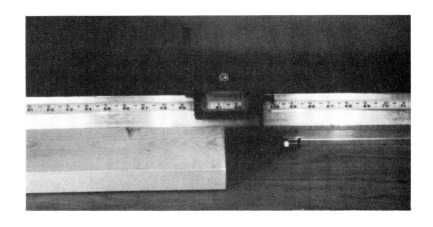

Illus. 727. This stop helps make saw setup efficient and accurate. Repetitive cutting is much faster.

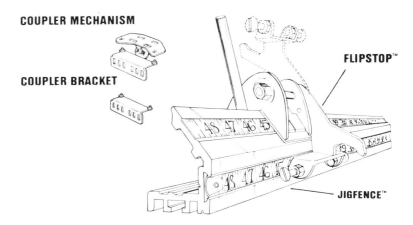

Illus. 728. These commercial accessories were used to make the Sawbuck extension table.

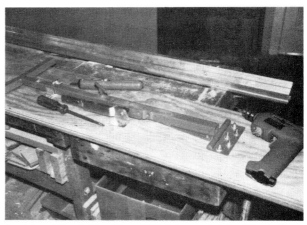

Illus. 729 (above left). Fasten the coupler mechanism to the underside of the table. Illus. 730 (above right). Attach the telescoping leg assembly to the table at the end opposite the coupling mechanism.

Specialty and Stationary Circular Saws

10
Specialty and Stationary Circular Saws

Some portable circular saws are not easy to classify. They may have a specialized purpose or an operating design that is unlike any other. This chapter is devoted to these circular saws and stationary circular saws.

Just as consumers have bought smaller cars, they are also buying smaller or "portable" stationary tools. Smaller tools allow the woodworker to make better use of the workshop; these tools can be stored when not in use, which allows the woodworker to use floor space for assembly of the project after the sawing is complete.

Note: Due to the unique design and features of the saws presented in this chapter, a complete list of safety procedures cannot be included. You are urged to review all safety precautions in this book and to read the owner's manual before operating the machines discussed in this chapter.

Special-Purpose Saws

One special-purpose saw is the Casey Hand Tool or saw. This device is used to cut straight-line openings in wood doors and panels, and with a special setup can cut dadoes in cabinet sides (Illus. 731). It uses a 4¾-inch-diameter dado head (Illus. 732). Specialty guides can be used with the dado head or saw blade; these guides control the path of the blade, and can also be adjusted for angular cuts.

The Casey tool is favored by small-to-medium cabinet and millwork shops. It is versatile and does not consume as much floor space as stationary equipment. It is also much less expensive than stationary equipment.

When using the Casey Hand Tool, be sure to observe the same safety practices you would with any portable circular saw. As with any other new tool, you should read and understand the owner's manual before using the tool.

Another special-purpose saw is the biscuit joiner (Illus. 733). While this tool uses a saw blade, it is better known for joining stock than for sawing it. The biscuit joiner is a European tool that makes assembly of flat panels accurate and easy (Illus. 734). It can be found in almost any cabinet shop where large amounts of plywood or other panel stock are used.

The secret of this system is the elliptical or football-

Illus. 731. The Casey Electric Hand Tool can cut dadoes in cabinet sides.

Illus. 732. The Casey Tool uses a 4¾-inch-diameter dado head. A specialty guide controls the path of the dado.

Illus. 733. The biscuit joiner is a specialty sawing machine. It is used to make joints between parts.

Illus. 734. Flat panels can be joined easily with biscuit joinery.

shaped biscuit (Illus. 735) that fits into a kerf cut by the saw blade. The biscuits come in three sizes for various types and sizes of work.

To accommodate the different sizes of biscuits, the blade depth (Illus. 736) is controlled by a stop mechanism (Illus. 737). Once adjusted, the stop automatically exposes the correct amount of blade. The biscuit joiner can also be used for other purposes such as making splines. In addition, hinges, brackets, and knockdown fasteners have been designed for use in the kerf cut by the saw.

There are at least four types of biscuit joiners on the market today. All have essentially the same controls and adjustments.

The biscuit joiner is a special-purpose saw, but it still has a blade that will cut anything in its path. Be sure to clamp the work; this way, you can keep both hands on the saw and out of the blade's path. Read and observe the safety practices offered by the manufacturer in the owner's manual.

While chop-stroke and pull-stroke mitring machines have been discussed in earlier chapters, one mitring machine is unique. It is a combination chop-stroke and push-stroke mitring machine (Illus. 738). This DeWalt machine is sold under the trade name Crosscutter™; it is also capable of compound-mitre cuts (Illus. 739). Mitre settings on this machine are made in the same way as they are on other compound-

Illus. 735. Insert the elliptical-shaped biscuits into the cut made by the biscuit joiner.

Illus. 736. Blade depth is limited by a stop. The amount of blade exposed determines the size of the biscuit used.

Illus. 737. Turn the stop mechanism to control blade depth for cutting joinery.

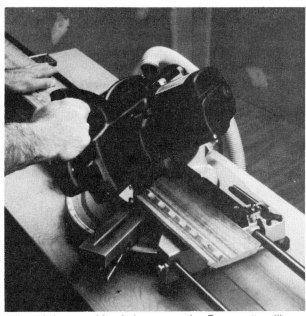

Illus. 738 (above left). This combination chop-stroke and push-stroke mitring machine is known as the Crosscutter. Illus. 739 (above right). The Crosscutter cuts compound mitres on crown moulding well.

mitre cutting machines. Consult the owner's manual for specifics.

The Crosscutter is capable of relatively wide crosscuts (Illus. 740) and mitres (Illus. 741). It comes equipped with a work-clamping mechanism and dust-collection accessories. These accessories make any operation safer and easier.

Machines of Unique Design

There are several stationary machines that can be categorized as being both portable and unique in design. They are: the push–pull saw, the Versatile Saw™, the Norsaw™, and the Flip Saw. These machines all resemble table saws, but have unique features.

The Erika™, by Mafell, is probably the best-known push–pull-type saw. It is manufactured in Germany. This saw is used like any conventional table saw. In addition, it has a pull-stroke blade. The motor travels on a pair of rods; this allows the blade to be pulled through a piece of work that is held stationary (Illus. 742). This feature allows it to be used in the same way as a radial-arm saw. To do mitre work, turn the work to the desired angle and pull the blade through it. The blade can be tilted for a compound mitre.

Mitres can also be cut in the normal way using the mitre gauge. The pull stroke is useful for large work or

Illus. 740. Here the crosscut is being made with the work clamped. Note the push stroke. The hoses collect and remove dust.

Illus. 741. Cut the face mitre on the push stroke. Always keep your hands clear of the blade's path.

Illus. 742. The blade can be pulled through the work while it remains stationary.

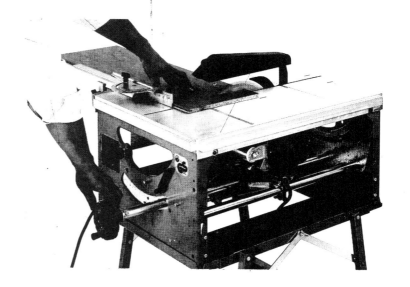

heavy timbers. The Erika is designed for work on the job, but may also be used in the shop. The basic machine weighs about 82 pounds, so one person can transport the machine.

The Versatile Saw™ is designed like a 10- or 12-inch table saw. In addition to the standard rip cuts and crosscuts a typical table saw performs, the Versatile Saw will perform cutoff work. The arbor and saw blade travel the length of the table.

By unlocking the arbor and pulling the T handle on the front of the guard, you can make a crosscut. Hold the stock against the mitre gauge while pulling the blade through the work; this allows the machine to crosscut in the same manner as a radial-arm saw. You can cut end mitres, face mitres, and compound mitres by pulling the arbor into the work.

The guard travels with the blade during a crosscut so that it is guarded throughout the operation. You can mount extension wings for more table-saw versatility, or build a portable cutoff bench for cutoff work.

The Versatile Saw was designed in 1947 in Henniker, New Hampshire, where it is still being produced. Before using the Versatile Saw, consult the owner's manual for the safe and correct operating procedures.

The Norsaw™ is similar to a push–pull saw (Illus. 743). This machine has been in use for about 30 years, but distribution has been limited, until recently, to Scandinavia. The Norsaw has many galvanized parts which make it an all-weather saw. The blade is lifted vertically through the work instead of being pulled through the work. The machine is similar to an inverted motorized mitre box.

The blade and motor are suspended from a turntable. Turn the turntable 90 degrees when changing from a rip cut to a crosscut. You can make a crosscut by pulling the work into the blade or by lifting the blade up through the work. You can cut mitres by turning the turntable to the desired angle and lifting the blade up through the work.

The Norsaw has an extension table that supports the work and acts as a fence when ripping. The blade can be tilted for bevel or chamfer cuts in the rip mode.

You can cut dadoes and rabbets by setting the blade to the desired height and making a series of rip cuts or crosscuts. You can also make pattern cuts by using a jig or fixture to control the work.

There are three models of the Norsaw: the models 805, 1203, and 1603. The 805 has an 8-inch blade, the 1203 a 12 inch, and the 1603 a 16-inch blade. The 1203 weighs 160 pounds, and the 1603 weighs 285 pounds. These machines require two persons for transport unless a roller stand is used.

The Flip Saw is manufactured in France by the Elu Machinery Company (Illus. 744). The unique feature of this saw is that it can be converted from a table saw to a motorized mitre box if you flip the table upside down. The saw weighs 87 pounds with its legs, so transport by one person is possible. Elu makes a complete line of accessories, including a sliding table, table-extension rods, a work stop for controlling part length, and a dust-extraction kit for controlling dust.

The Flip Saw uses a 10-inch blade. It is driven by a 14-ampere induction motor. Induction motors have no brushes and require little maintenance. The motor output is rated at 1.5 horsepower. The Flip Saw is designed to cut metal tubing as well as wood.

You can operate the Flip Saw the same way as any other table saw when you set it up in that mode.

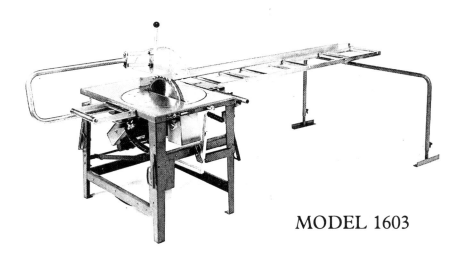

Illus. 743. The Norsaw is similar to a push–pull saw.

MODEL 1603

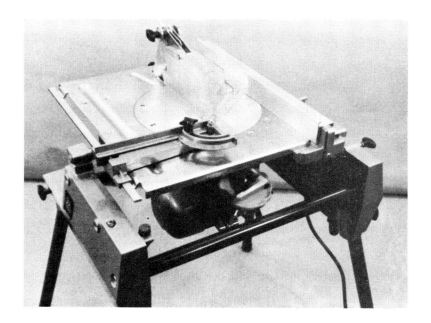

Illus. 744. The Flip Saw looks like a table saw, but can be flipped over into a motorized mitre box.

Observe the appropriate safe procedures when operating it. The blade can be tilted for mitre or compound-mitre cuts. You can cut end and face mitres using the mitre gauge to control stock (Illus. 745 and 746), and edge mitres using the fence to control stock (Illus. 747).

You can make rip cuts (Illus. 748) and crosscuts (Illus. 749–751) the same way on the Flip Saw as you would on a conventional table saw. Be sure to observe safety operating procedures.

To convert the Flip Saw to a motorized mitre box, lock the blade in the upright position and remove the guard (Illus. 752). After you have adjusted the blade height to maximum height, release the table-locking pin (Illus. 753) and flip the table into the mitre position (Illus. 754–756).

You can use the table extension/work supports with

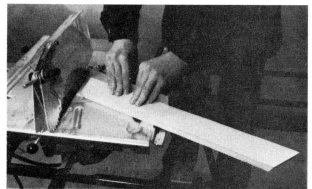

Illus. 746. To cut a compound mitre, tilt the blade and turn the mitre gauge.

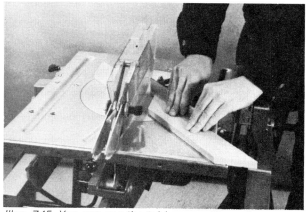

Illus. 745. You can use the table saw with a mitre gauge to cut a mitre.

Illus. 747. To make edge mitres, tilt the blade and control the stock with the fence.

Illus. 748. You can make rip cuts the conventional way on the table-saw side of the Flip Saw.

Illus. 749. The crosscut is also made the conventional way on the table-saw side of the Flip Saw. Stock is controlled by the mitre gauge.

Illus. 750. An extension table supports long, wide pieces on the table-saw or motorized mitre box side.

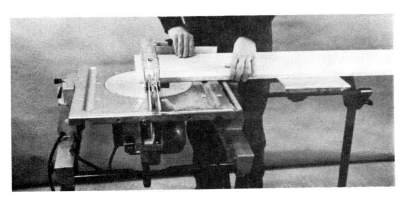

Illus. 751. The extension table supports a long board during a table-saw crosscut.

Illus. 752. To flip the saw to the motorized mitre box, remove the guard and expose the blade fully. Disconnect the power when flipping the saw.

Illus. 753. Next, release the pin. Hold the table securely when doing this.

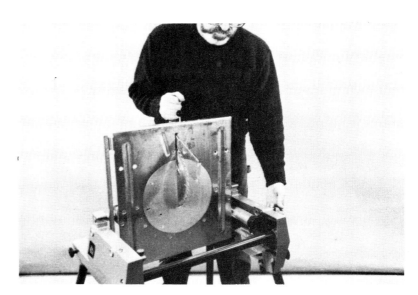

Illus. 754. Turn the table on its axle.

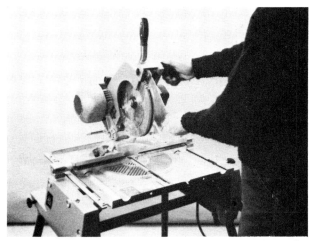

Illus. 755 (above left). Release and store the splitter. Illus. 756 (above right). The Flip Saw is now ready for mitre work.

a stop to control part length. Make cuts the same way you would make them with any other motorized mitre box (Illus. 757–765). Be sure to observe the safe operating procedures presented in Chapter 8.

Portable Stationary Saws

Table Saws Some table saws can be considered bench-top or portable saws. These saws typically use an 8-inch blade. Some come equipped with a stand, while others come designed to clamp to a bench or sawhorse. Most portable table saws are motorized rather than motor-driven (belt drive); this makes the unit more compact.

Bench or portable table saws usually have a smaller table; this can limit control when you are cutting large pieces (Illus. 766). Generally, there is little table in front of the blade when it is at full height. With practice and the correct accessories, you'll find that the portable table saw can be extremely useful on the job or in the small workshop.

Make rips by setting the fence and feeding the stock into the blade (Illus. 767). Narrow pieces should be controlled with a push stick. Use a guard whenever possible (Illus. 768).

Make crosscuts using the mitre gauge to control stock (Illus. 769). Some mitres are made by tilting the blade (Illus. 770 and 771). Make compound mitres by tilting both the blade and the mitre gauge.

The power rating of portable table saws varies according to make. Avoid overworking a portable table saw. Take light cuts and feed the work only as fast as the blade can cut it. Be sure the saw is level when you set it up. Any rocking or vibration will make the job unsafe and produce a substandard cut.

Illus. 757. A stop on the extension table controls the length of the stock when you are crosscutting.

Illus. 758. Position the stock for a left-mitre cut.

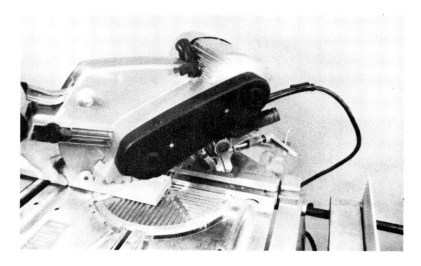

Illus. 759. Make the mitre cut using conventional mitring practices. These practices are discussed in Chapter 8.

Illus. 760. Set up the left-mitre cut using a stop to control stock length.

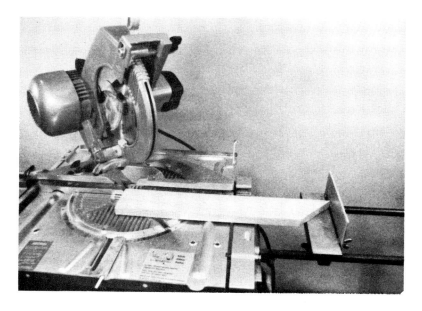

Illus. 761. The Flip Saw makes mitre cuts that are clean and efficient.

Illus. 762. End mitres can be cut on stock of moderate width.

Illus. 763. Use the Flip Saw as you would any other motorized mitre box.

Illus. 764. The arm can be tilted on the Flip Saw. Lock it securely to the post when tilting it.

Illus. 765. When the arm is tilted, a compound mitre can be cut.

Illus. 766. The distance from the front of the table to the front of the blade (at full height) determines the size of the stock that can be cut. The greater this distance, the better control you can have over the stock.

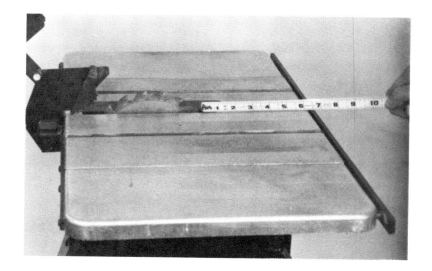

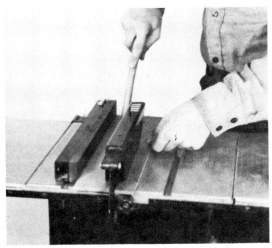

Illus. 767. When ripping, guide the work into the blade at uniform speed. If the blade slows down, you are feeding too fast. Burned edges can mean the blade is dull or that you are feeding too slowly.

Illus. 768. This extension supports long rips. Note that the guard can be used with this shop accessory.

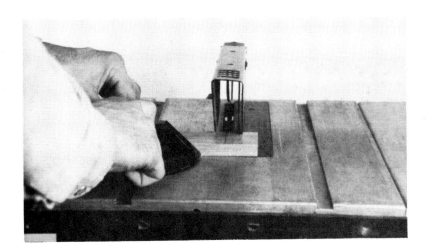

Illus. 769. Advance the mitre gauge into the blade at a moderate speed when crosscutting. Feeding too slowly wastes time, and feeding too quickly increases tear-out. Keep the good or exposed face of your work up so that any tear-out occurs on the back or unexposed side of the work.

Illus. 770. This edge mitre is made by tilting the blade. It is a rip-type cut. Note that a push stick is being used to control the stock.

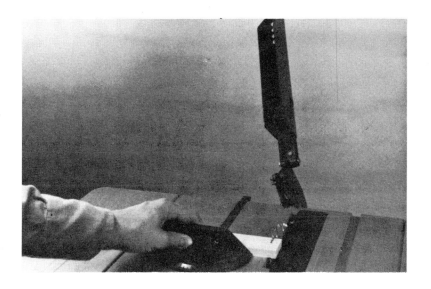

Illus. 771. This end mitre is made with the blade tilted. It is a crosscut. Note that the guard has been used.

When transporting a portable table saw, handle it carefully (Illus. 772–776). Abusive handling can affect the adjustment of the saw. It is impossible to get a quality cut when the saw is out of adjustment. Check the saw periodically to be sure it is adjusted correctly.

For more information concerning table-saw adjustment and use, consult *Table Saw Techniques* (Sterling Publishing Company, Two Park Avenue, New York, New York 10016).

Radial-Arm Saws Portable radial-arm saws typically use an 8–10 inch blade. Some come equipped with a stand, while others are designed to clamp to a bench top (Illus. 777–779). The saw shown in Illus. 780 is a 10-inch machine. The stand is the two folding sawhorses on which the saw is resting.

The handles attached to the saw (Illus. 781) are clamped to the sawhorses; these handles allow the saw to be transported easily by two people. This saw is too heavy to be transported by one person. The handles also protect the saw from getting out of adjustment.

When a radial-arm saw is transported by the table, base or column, the parts are likely to shift; this can affect the accuracy or the quality of the cut. If you attach the handles, you will ensure that the relationship of the saw parts will not be affected during normal transportation.

The Inca radial-arm saw uses a 9-inch blade. The stand and saw are assembled from square tubing and connectors. This saw can be transported by one person because it can be broken down into smaller units (Illus. 782–784). The carriage travels on a piece of square tubing. To use it for ripping, turn it 90 degrees (Illus. 785). To make mitre cuts, turn the table (Illus. 786).

The Ryobi radial-arm saw is an 8¼-inch machine. It

Illus. 772. This small base can be easily moved to the work site.

Illus. 773. This small saw can be easily moved by one person. It can be attached to the base at the work site.

Illus. 774. The wooden base holds the cord inside the saw and protects the bolts from damage.

Illus. 775 (above left). The bolts are used to secure the saw to the base. Illus. 776 (above right). Attach the guard after setting up the saw.

Illus. 777 (above left). This stand was designed for a motorized mitre box. It can also be used with some radial arm saws.
Illus. 778 (above right). The radial arm saw has been clamped to the top of this stand.

Illus. 779. The accessories offered with the stand are used with stock support.

Illus. 780. The handles attached to this saw make it easier to transport and allow it to be clamped to a pair of sawhorses.

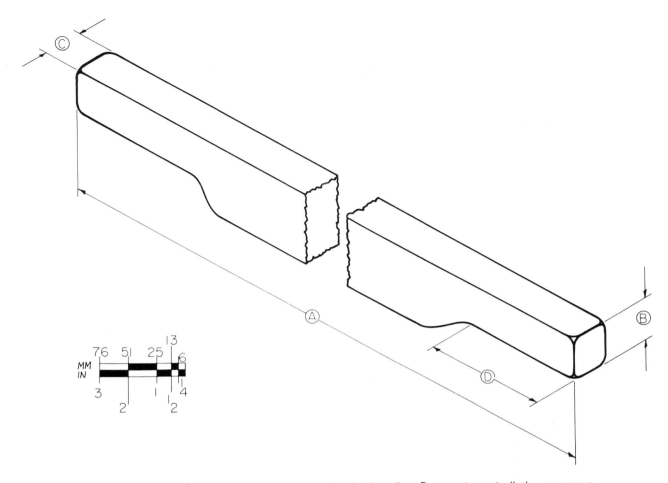

Illus. 781. Use this drawing to develop a pattern for shaping the handles. Be sure to rout all sharp corners.

Illus. 782. The Inca saw has a base and frame made of square tubing.

Illus. 783. The short overarm has been installed here. A wider crosscut can be made with the longer overarm.

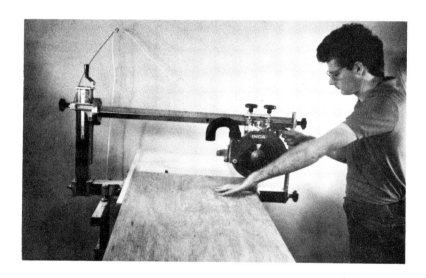

Illus. 784. The long overarm allows a crosscut of over 24". Note that the front support bar is attached to the arm for rigidity.

Illus. 785 (above left). This saw is ready for ripping. Note that the splitter has been positioned. Illus. 786 (above right). Turn the table for angular cuts. The overarm remains stationary.

is light enough to be transported by one person, and powerful enough to cut to its full capacity. The saw does not come equipped with a stand, but it can be clamped to most motorized mitre-box stands.

The Ryobi radial-arm saw uses the standard radial-arm-saw accessories and cutters (Illus. 787–790). You can make crosscuts in the conventional way (Illus. 791). Mitres can be cut by turning the arm or tilting the blade. Compound mitres are cut by turning the arm and tilting the blade.

You can make rip cuts by feeding the stock into the saw. Anti-kickback pawls and some kind of hold-down should be used when ripping (Illus. 792 and 793); these devices reduce the chance of kickback. Always keep your hands a safe distance from the blade. Be sure to use the guard for all cuts.

Make sure the saw is anchored and clamped securely before cutting. Check all clamps and locks before making any cut. Make adjustments with the power disconnected; you will have to make many on the radial-arm saw. A loose clamping mechanism could affect the quality of the cut or make the cut unsafe.

For more information concerning radial-arm-saw adjustment and use, consult *Radial Arm Saw Techniques* (Sterling Publishing Company, Two Park Avenue, New York, New York 10016).

Illus. 787. The motor has a router collet on its opposite end. It turns high rpm's to support router work.

Illus. 788. Routing on this radial arm saw is controlled by the fence and table.

Illus. 789. A wobble-type dado head is being used here to cut dadoes.

Illus. 790. A shaper head is being used here to shape the face of a piece of stock. Portable radial arm saws can do the work of stationary saws.

Illus. 791. This portable radial arm saw makes crosscuts in the conventional manner.

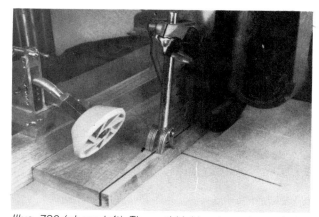
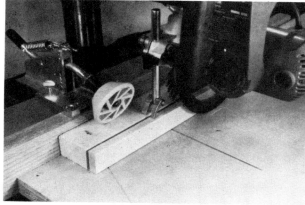

Illus. 792 (above left). The anti-kickback pawls are being used to minimize the chance of kickback while the operator is ripping. Illus. 793 (above right). The portable radial arm saw will rip a piece of 1½-inch stock even at a 45-degree angle.

Projects

11

Accessories for Your Portable Circular Saws

The projects in this chapter consist of accessories for your portable circular saws. These projects have been designed to help you become proficient with your saws. It is important that you digest the material in the first 10 chapters before you begin building the projects here. If you begin a project without a full understanding of how to build it, you could waste material or injure yourself. Plan projects carefully and review unfamiliar operations before beginning.

Tips for Building Projects

When you start a project, consider the following tips. These tips will help you do a better job with fewer mistakes.
1. *Study the Plans Carefully.* Check the dimensions or the scale on the plan to be sure that all parts will be cut correctly. Make allowances for joints or trimming on all parts. Some plans are dimensioned for a particular saw. You may have to modify those plans to fit your saw.
2. *Develop Detail or Auxiliary Sketches When Needed.* When you are modifying plans, or an assembly is complex, develop an auxiliary sketch that can simplify the job.
3. *Develop a Bill of Materials.* A bill of materials is a list of all parts needed to build the project. The list includes the dimensions of the parts (thickness, width, and length). The bill of materials makes cutting the parts more efficient and orderly.
4. *Write a Plan of Procedure.* The plan of procedure is a list of steps you should follow to build a project. The list is an orderly series of events. Many times one part determines the layout on the next. The plan of procedure controls the sequence of events.
5. *Think Before Making Any Cuts.* Plan your cutting to reduce the chance of error. When cutting parts, cut the longest parts first. Any parts cut too short can then be used for shorter parts for the bill of materials.

Before making a cut of your work, test the setup in scrap. Make sure that it is correct before you cut the work.
6. *Check the Fit of all Parts Before Assembly.* Fit the parts together dry before gluing. Tight fits or incorrect fits can mean trouble once the glue has been applied. It is much easier to trim parts or make adjustments when there is no glue on the parts.

Projects

Saw Guides Many times it becomes necessary to make a very straight or square cut with a portable circular saw. While commercial devices are available to help you make accurate cuts, you can also make your own.

When you are crosscutting, a square corner of a sheet of plywood makes an excellent guide (Illus. 794). The base of the saw rides along the edge of the plywood (Illus. 795). A cleat on the bottom of the piece of plywood holds the guide perpendicular to the work.

Make the jig by finding a scrap of plywood with the factory corner intact. Mark the square factory corner, and cut out a piece approximately 8 × 12 inches. Glue and nail a cleat to the bottom of the plywood corner (Illus. 796). The cleat must line up perfectly with the factory edge (Illus. 797). This ensures that the guide will cut a square corner.

If you want to make a guide that cuts long, straight lines (Illus. 798), a different procedure is required. See pages 60–67 and Illus. 153–156 for complete details.

Sawhorses The most important work-supporting device used with the portable circular saw is the sawhorse. The sawhorse holds work at a comfortable height for sawing. It also provides a clamping surface

Illus. 794. The factory corner from a sheet of plywood makes an excellent crosscutting guide.

Illus. 795. The saw base rides along the edge of the plywood. This controls the blade's path. A cleat on the bottom of the guide butts against the front edge of the work.

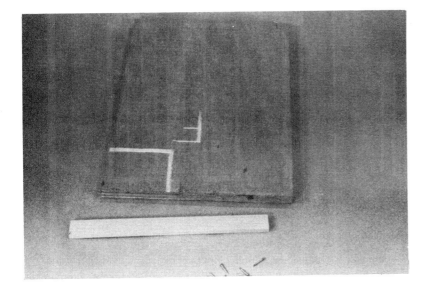

Illus. 796. Glue and nail a wood cleat to the edge of the plywood cut-off.

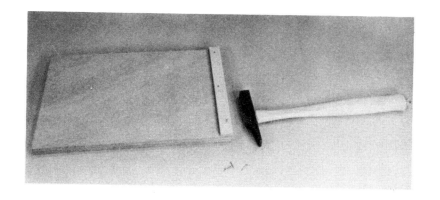

Illus. 797. To obtain square cuts, make sure that the cleat lines up perfectly with the factory edge.

Illus. 798. A guide like this one enables you to cut long, straight lines. Construction details are presented in Chapter 4 (pages 60 and 63–65).

for the work; this makes sawing safer and more accurate. Clamping devices can also be integrated into the design of your sawhorse (Illus. 799).

Two different-sized sawhorses are presented in Illus. 800. They differ in height and width, but they both have many uses. On the job, one coworker referred to the small horse as a saw pony; the work surface on this sawhorse is also smaller (Illus. 801), but it can be modified according to your individual needs.

Study the plans for the two sawhorses (Illus. 802 and 803). Develop a bill of materials and select your stock; look it over carefully. Try to find true, clear stock for your horses; they will look better, perform better, and be stronger.

Most woodworkers build sawhorses in pairs. My preference is to build them in sets of three or four; this way, when I handle large panels or long timbers, they will have better support and will be easier to cut. Two horses can be on either side of the cut or the cut can be made over the third sawhorse.

All angles on the sawhorse are 15 degrees (Illus. 804). Set your sliding T bevel to this angle and lock the setting. Use this angle setting to adjust the angle between the base and the saw blade. Lay out the sawhorse work surface, and bevel-rip the edges (Illus. 805). Tip the top slightly wide so that the edges can be planed or sanded smooth.

After you have ripped the tops, you can cut the legs according to the plan (Illus. 806). Illus. 219–222 and 226–230, on pages 87–90, depict the entire cutting sequence.

Use the legs to lay out the notches that must be cut in the top (Illus. 807). Set the blade depth to the thickness of the legs and make the shoulder cuts. If you have a Thorsness or Laser blade, you can saw-out the middle with a back-and-forth motion. If not, you can make a series of cuts; these cuts will establish the bottom of the notch (Illus. 808). The remaining wood can be chiselled away (Illus. 809). When you have removed most of the stock, use the chisel carefully to make the bottom of the notch smooth (Illus. 810).

Now glue and nail the legs into the notches. Gal-

Illus. 799. The clamping mechanism holding the work on this sawhorse was integrated into the sawhorse design.

Illus. 800. The two sawhorses are of different heights and widths. Select the one that best suits the work you do.

Illus. 801. The work surface on the small sawhorse is proportionately smaller. This can be modified according to individual need. The slope on the sides of the horse is 15 degrees.

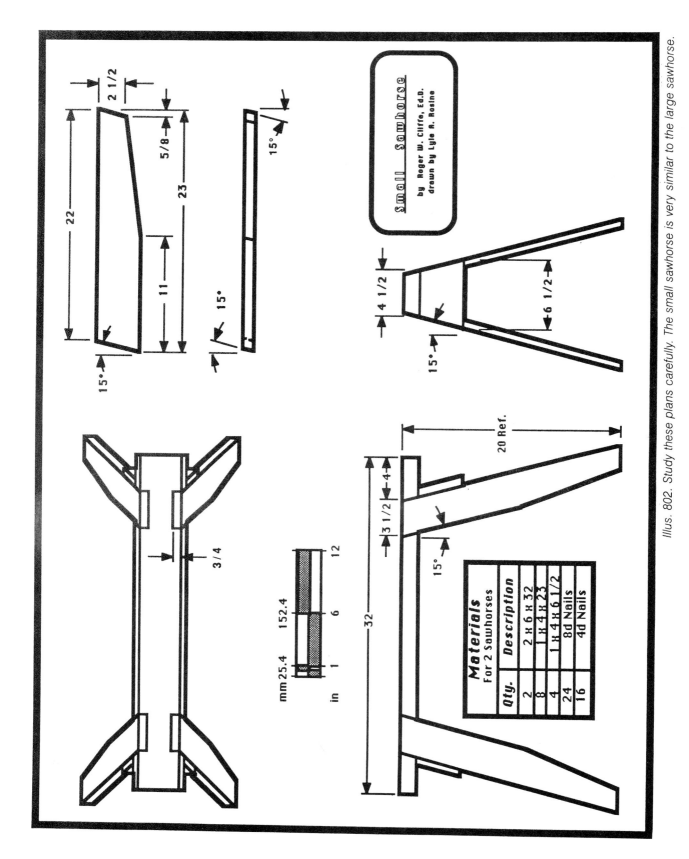

Illus. 802. Study these plans carefully. The small sawhorse is very similar to the large sawhorse.

Large Sawhorse

Drawn for: Roger Cliffe, Ed. D. *Drawn by:* Lyle A. Rosine

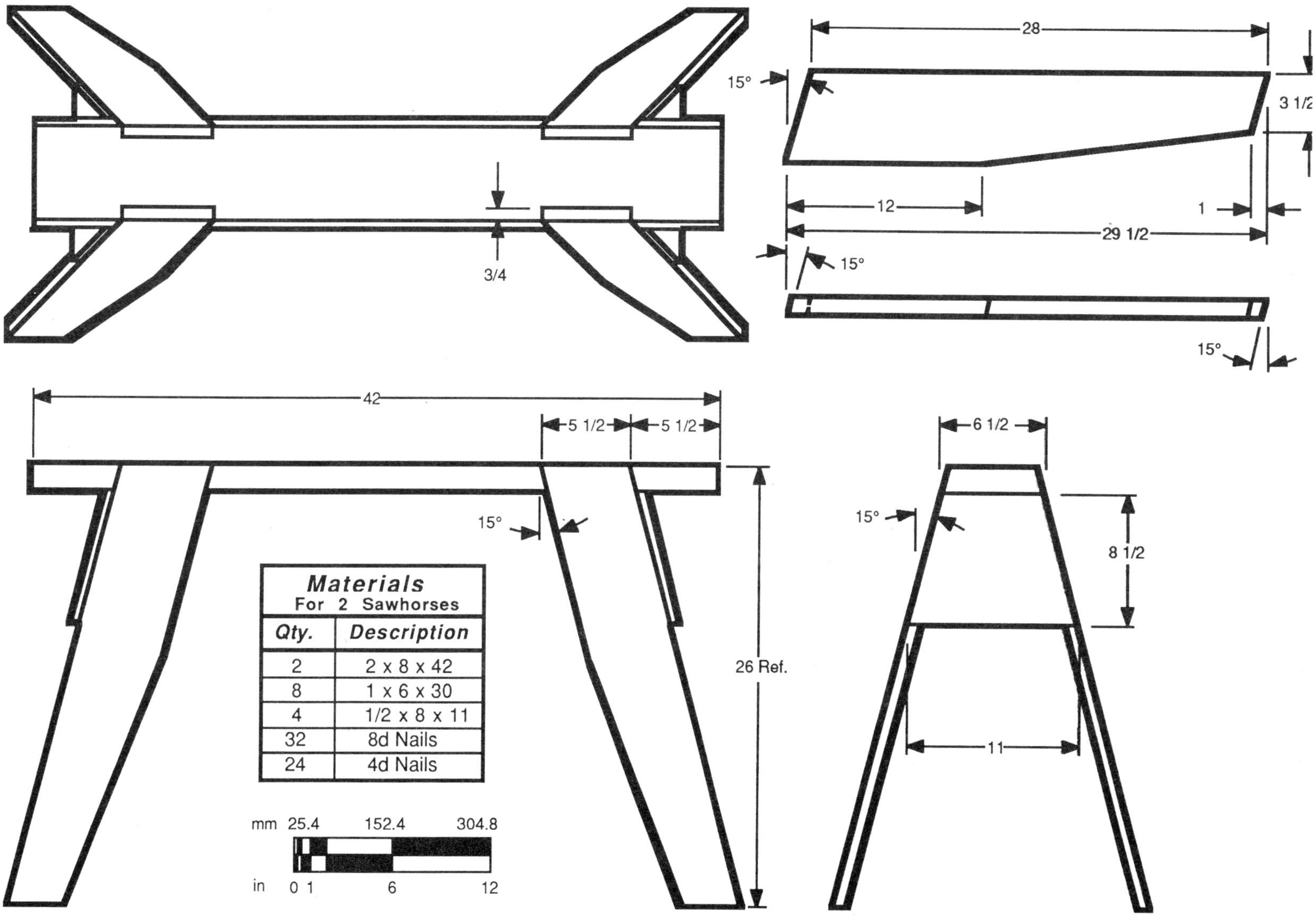

Materials For 2 Sawhorses	
Qty.	Description
2	2 x 8 x 42
8	1 x 6 x 30
4	1/2 x 8 x 11
32	8d Nails
24	4d Nails

Illus. 803. After you study the plans, decide on which sawhorse you want to build, and then develop a bill of materials.

Illus. 804 (above left). All angles on this sawhorse are 15 degrees. Set your sliding T bevel to that angle and lock the setting. Use it for stock layout and saw setup. Illus. 805 (above right). After the saw has been set at 15 degrees, bevel-rip the top to size. Note that the rip fence has been used to control the saw.

Illus. 806. Study the plans and then cut the legs. Before you lay out the taper, remember that there are right and left legs.

Illus. 807. Use the leg to lay out the top for notching.

Illus. 808. You can make a series of saw cuts as deep as the leg is thick to form the notch.

Illus. 809. Use a chisel to clear away the stock in the notch.

Illus. 810. Pare away the stock carefully when you get near the bottom of the notch. The bottom of the notch should be very smooth.

Illus. 811. Glue and nail the legs into the notches. Avoid splitting the wood when nailing. This weakens the sawhorse.

vanized nails (Illus. 811) resist rust and have greater holding power. Use a nail about two inches long. The nail diameter should be minimal so that the leg will not split when you drive the nail in. There is little holding power when the nail splits the wood.

If you want to put a lower shelf on the sawhorse, you will have to nail some cleats to the middle of the legs (Illus. 812). Lay out their positions carefully, and then glue and nail them to the legs (Illus. 813). After you have positioned all the cleats, glue and nail a plywood shelf in position (Illus. 814).

Now install the end braces. The edge of the brace that goes adjacent to the top has to be bevelled 15 degrees (Illus. 815). Use the legs to lay out the inclined edges of the brace (Illus. 816). Clamp the braces securely (Illus. 817) and saw them carefully (Illus. 818). If they are slightly oversize, they can be sanded flush after assembly (Illus. 819).

The end braces can be made of plywood or solid stock. The braces are wider when the sawhorse has a shelf. Glue and nail the braces in position (Illus. 820). A nail set may be helpful when driving nails near the top (Illus. 821).

After the glue cures, use a belt sander to even up the top and smooth the braces flush with the legs (Illus. 822). The sawhorse is now ready for extras such as a rip slot or a clamping mechanism, if desired. A rip slot provides blade clearance when you are ripping small pieces. The blade can freely turn in the rip slot without damaging the sawhorse (Illus. 823). If you build a set of sawhorses, you should have a rip slot for those small rip jobs.

If you want to cut a rip slot, lay out two centers for a hole about two inches in diameter (Illus. 824). Cut the holes with a hole saw or other device (Illus. 825). Connect the holes with a straightedge (Illus. 826). Set up the plunge saw to cut the lines between the holes (Illus. 827). Cut inside the layout lines (Illus. 828). Be

Illus. 812. If you want a lower shelf, cleats must be nailed to the legs. The cleats support the plywood shelf.

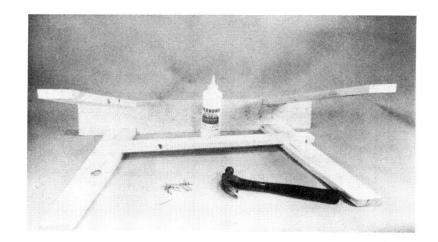

Illus. 813. Use glue and nails to secure the cleats after their position has been laid out.

Illus. 814. Slide the plywood shelf in from the end and glue and nail it in position.

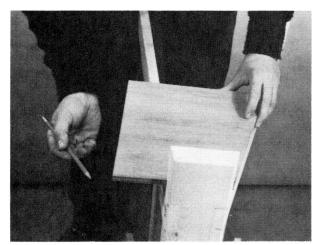

Illus. 815 (above left). The upper edge of the end brace must have a 15-degree bevel to fit correctly. Illus. 816 (upper right). Use the legs to lay out the brace accurately.

Illus. 817. Clamp the brace securely for sawing. If you use a plunge saw, adjust the blade depth before cutting.

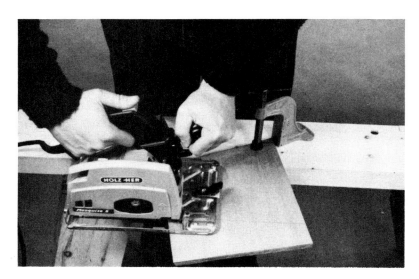

Illus. 818. Saw the braces carefully. Try to leave the layout line. It is better that the braces be a little big.

Illus. 819. If the braces are slightly oversize, they can be sanded flush after assembly.

Illus. 820 (above left). Glue and nail the brace in position. This plywood brace extends to the bottom of the shelf support.
Illus. 821 (above right). Use a nail set to drive the nails near the underside of the top. A solid wood brace was used on this sawhorse. It did not have a lower shelf.

Illus. 822. Even up the top and the braces with a sander after the horse is assembled and the glue cures.

Illus. 823. The saw blade can turn freely in the rip slot without damaging the sawhorse.

Illus. 824. Begin the rip slot with two holes about 2 inches in diameter.

Illus. 825. A hole saw or spade bit works well for cutting the holes.

Illus. 826. Connect the holes with a straightedge.

careful not to saw beyond the holes (Illus. 829). A conventional saw can also be used to make this cut (Illus. 262 and 263). Cut away the remaining stock with a hand saw (Illus. 264); this will remove the waste stock and form the rip slot.

A clamping mechanism uses a conventional hardware device (Illus. 830) to hold the work while sawing. The bolt to which the clamp is secured drops below the top when it is not in use. Extra bolts are available that allow you to position the clamp in several different places.

Decide where you wish to locate the clamps and mark the bolt centers. Drill a hole slightly larger than the bolt head and slightly deeper than the thickness of the bolt head (Illus. 831). Then drill a through hole slightly larger than the bolt diameter (Illus. 832) using

Illus. 827. Set up a conventional saw or plunge saw to cut out the rip slot.

Illus. 828. Cut inside the layout lines. Keep the saw on course and do not back up. Simply start the cut again.

Illus. 829. Do not cut beyond the hole. Saw carefully and cut away the remaining stock with a hand saw.

Illus. 830. This commercial clamping device can be used on a sawhorse to hold stock securely.

Illus. 831. Locate the clamping position and drill a hole slightly larger than the bolt head. The hole depth should be slightly greater than the thickness of the bolt.

Illus. 832. Drill a through hole on the same center. This should be slightly larger than the threaded diameter.

the same center point. Then push the bolt through the hole, adjust it to accommodate the clamp (Illus. 833), and fasten a washer and two nuts to the bolt.

The sawhorses may not sit evenly on all four legs after assembly; to adjust them, saw the two longer legs a second time. If you have made a pair of sawhorses, start with the shortest horse. Shim the short legs until the horse sits evenly on a true surface (Illus. 834). Mark the legs with a spacer that is thicker than the shim wedged under the short legs (Illus. 835). Saw the legs along the lines marked with the spacer (Illus. 836). After shimming the short sawhorse, move it closer to the taller sawhorse. Spacers should be equal to the difference in sawhorse height (Illus. 837). Mark the legs accordingly (Illus. 838) and saw them accordingly.

Protect the sawhorses with one or two coats of penetrating oil. Apply additional coats if the sawhorses appear to dry out with use. Penetrating finishes are best because they do not flake or peel with exposure to the elements.

Illus. 833. Push the bolt through the hole and adjust it to accommodate the clamp.

Illus. 834. Shim the sawhorse until it sits properly on a true surface. If you are working on more than one sawhorse, begin with the shortest one.

Illus. 835. Mark all legs with a spacer greater than the shim thickness. Trim the legs accordingly.

Illus. 836. Saw along the line marked on the legs. Clamp the sawhorse for safe and accurate sawing.

Illus. 837. To make all horses equal in height, put a shim on top of the shortest one. It should be the same thickness as the differences in height.

Illus. 838. Use that shim to mark the legs of the taller sawhorse. Cut on the waste side of the line.

Saw Bench This saw bench (Illus. 839) is ideal for supporting long pieces. It is also the ideal height for clamping stock with your knee (Illus. 840). If you have already built the sawhorses and plywood sawing jig, you can use those devices to saw the plywood.

The saw bench offers challenges such as sawing plywood, ripping thin strips, and cutting notches. These challenges can help you become a better woodworker and give you greater skill with your portable circular saw.

Study the plans (Illus. 841) and develop a bill of materials. Rip some thin strips to edge-band the plywood parts. Cut the notches in the ends to assemble the bench. Fasten the lower support with threaded cross dowels or other fasteners. Glue and nail the upper supports to the ends.

Cut the top larger than the base. This gives you many places to attach a clamp. Fasten the top to the base with glue and screws. The bench can be finished with a penetrating oil or paint. Paint tends to rub off on the work. If you use the bench more for picnics than for sawing, paint would probably be the best finish.

Illus. 839. This saw bench is ideal for supporting long pieces for sawing. It can also support about five people.

Illus. 840. The height is ideal for clamping stock with your knee. Note that the plywood has all been edge-banded. This makes the bench more attractive.

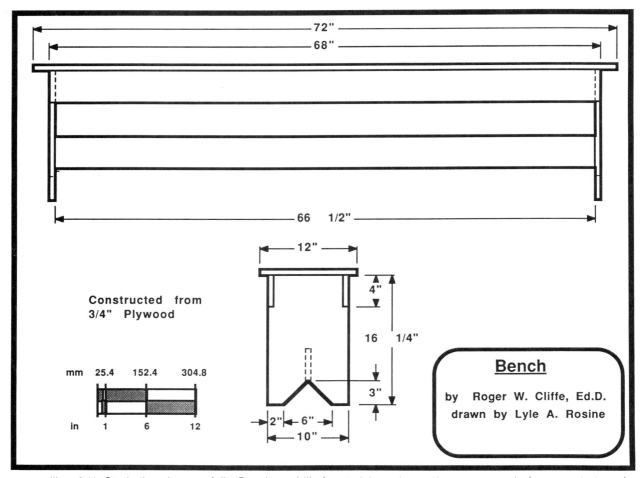

Illus. 841. Study the plan carefully. Develop a bill of materials and a cutting sequence before you start sawing.

Utility Box The utility box is a combination work support, tool box, and stool (Illus. 843). Carpenters like this device for trimming doors and windows. They also like the idea of storing tools under the work surface.

The utility box can be built out of sheet stock and/or solid stock. Study Illus. 842 and develop a bill of materials. After you cut all the parts to size, lay out the ends (Illus. 844). Start with the V-cut at the bottom (Illus. 845). This will require two cuts (Illus. 846). Do not cut past the layout line. You can cut the remaining stock with a hand saw.

Now cut the notch at the top of the side (Illus. 847). This will also take two cuts. Remove the remaining stock with a hand saw (Illus. 848). Clamp the side on edge (Illus. 849). Position the saw on the edge and cut on the layout line (Illus. 850). This cut forms the shoulder of the notch. The blade depth on the saw should be the same as the depth of the notch. Make a series of cuts (Illus. 851) until you come to the other shoulder (Illus. 852). The sides are now complete.

Join the bottom to the back with nails and glue. Turn the unit on edge and apply glue. Nail the side to the back and bottom (Illus. 853). Inspect the joints among the three parts (Illus. 854). Then nail the opposite side to the unit. Glue and nail the rails in position (Illus. 855).

Lay out the holes on the ends and drill them with a hole saw. Plunge-cut the opening between the holes (Illus. 856). Do not saw into the holes. Remove the remaining stock with a hand saw (Illus. 857).

Most woodworkers install a top that is exactly the size of the utility box (Illus. 858), but others prefer one that overlaps the sides (Illus. 859). A flush top makes the utility box easy to stack and carry. An overlapping top makes it possible to clamp parts to it from all edges.

Attach the top with glue and mild-steel wood screws. Mild-steel screws do less damage to carbide tips than hardened screws. If you accidentally hit a screw while sawing, it would be better if it were made of mild steel.

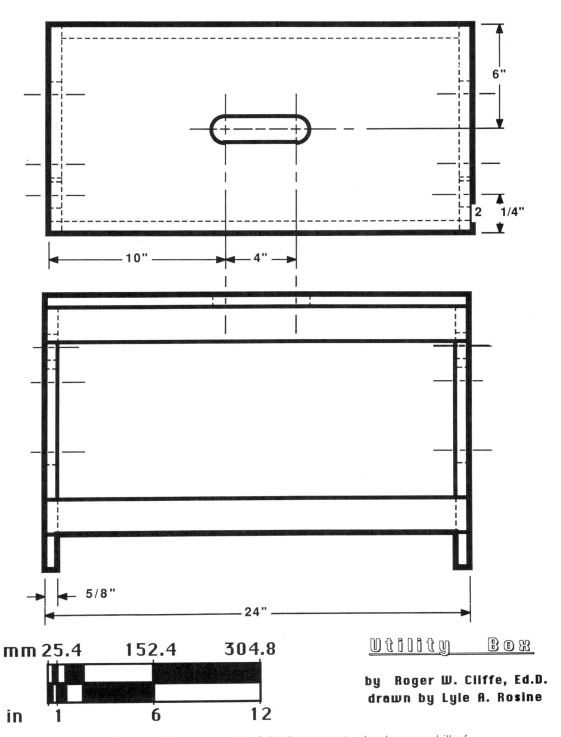

Illus. 842A. Use these drawings and the one on the following page to develop your bill of materials.

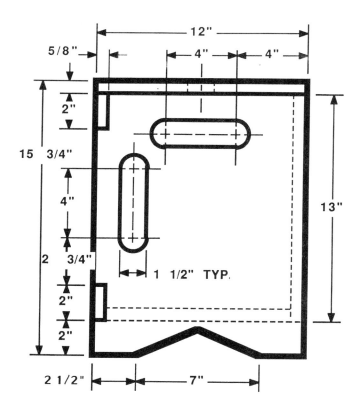

Illus. 842B.

Illus. 843. The utility box is a combination work support, tool box, and stool. This woodworker's helper is used by many carpenters.

Illus. 844. Lay out the ends after all the parts have been cut to size.

Illus. 845. Start cutting the end at the bottom.

Illus. 846. You have to make two cuts to form the V on the bottom of the sides.

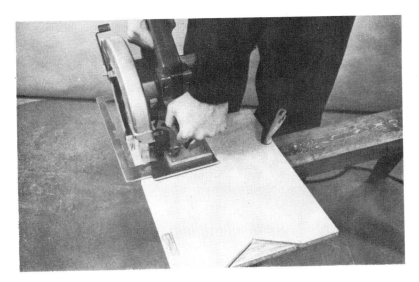

Illus. 847. Next, cut the notch at the top of the side. This will require two cuts.

Illus. 848. Use a hand saw to remove the cutaway portion of the side and bottom.

Illus. 849. Now clamp the side on edge so you can cut away the lower notch.

Illus. 850. Set the blade depth to the notch depth and saw along the waste side of the layout line.

Illus. 851. Make a series of cuts to remove the waste stock in the notch.

Illus. 852. Work carefully when you reach the layout line at the opposite end of the notch. Try not to remove too much stock.

Illus. 853. Glue and nail the side to the back and bottom. Line up the parts accurately before nailing.

Illus. 854. Inspect the fit between the mating parts. Apply clamps if necessary.

Illus. 855. Glue and nail the rails in position.

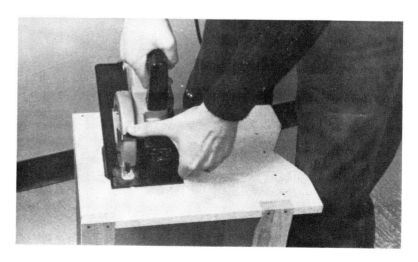

Illus. 856. Make two plunge cuts between the handle holes.

Illus. 857. Remove the remaining stock with a hand saw.

Illus. 858. A flush top makes the utility box easy to stack.

Illus. 859. An overlapping top provides more of a clamping surface, but is difficult to stack on its back.

Plywood-Sawing Jig Many woodworkers feel that sawing plywood in the field is one of the most difficult tasks with a portable circular saw because of the size of the panel and the difficulty of supporting it. The plywood-sawing jig (Illus. 860–862) supports the plywood panel and folds up for easy storage. Study the plan (Illus. 863) to determine overall size and to develop a bill of materials.

The plywood-sawing jig folds up with the help of a Roto-Hinge™ (Illus. 864). The hinge is glued into the two mating parts; it remains free and pivots on its center.

Start the plywood-sawing jig by selecting some clear, true framing members. Rip them carefully and crosscut them to length. Use the best stock you can find. Warped or defective stock will be difficult to assemble and may not be strong enough. Sand or plane the edges to smooth them.

Lay out the parts for drilling (Illus. 865). Be sure to mark the centers of the parts carefully (Illus. 866). Drill the holes to fit the Roto-Hinge. Make a test fit in scrap before drilling your workpieces (Illus. 867). Glue the hinge into the hole. Work carefully; do not put too much glue in the holes (Illus. 868). Fit the parts carefully and clamp them together (Illus. 869).

There are 20 hinges on the plywood-sawing jig, so you may want to divide the gluing job into subassemblies. The size of the finished jig allows it to rest on the top of either set of sawhorses; they were designed to be used together.

Apply some penetrating oil finish to the jig; it will protect the jig from the elements. The jig is now ready for use. When sawing the long way, make sure that the long members of the jig are up. When sawing across a panel, make sure that the short members are up. This provides clearance under the panel for the saw blade.

Illus. 860. This plywood-sawing jig spans two or more sawhorses and supports sheet stock for sawing.

Illus. 861. When sawing, place the sheet stock on the jig. When necessary, the stock can be clamped to the jig.

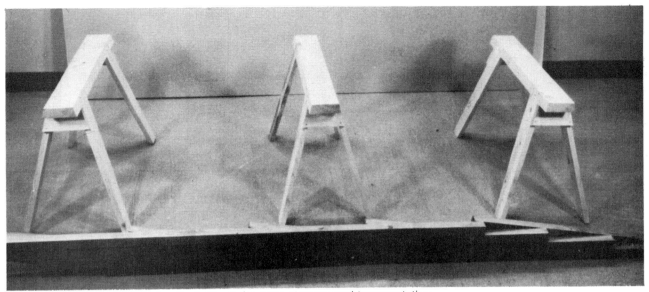

Illus. 862. The plywood-sawing jig folds up for easy storage and transportation.

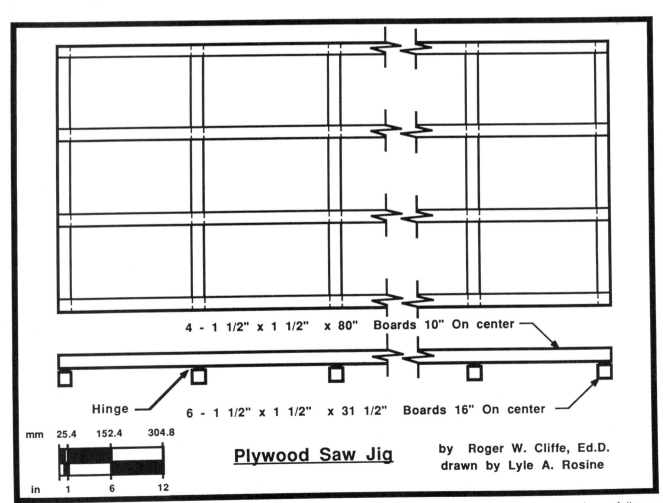

Illus. 863. Study the plans carefully before you begin. Work out your bill of materials and select your stock carefully.

305

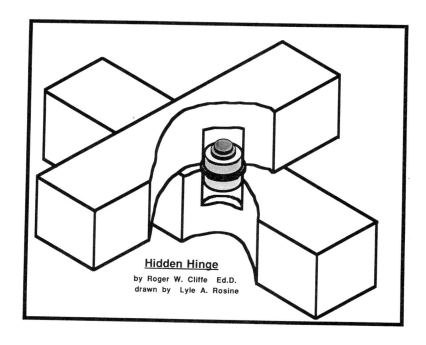

Illus. 864. Glue the Roto-Hinge into the mating parts. With it, the jig can be folded up easily for transportation and storage.

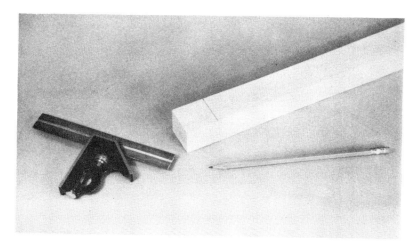

Illus. 865. Lay out the parts for drilling. The Roto-Hinge should fit snugly.

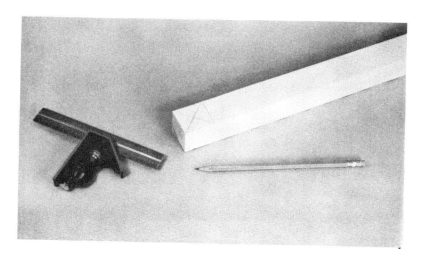

Illus. 866. To ensure that you place the hinges accurately, mark the centers carefully.

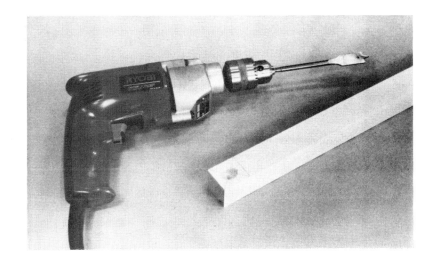

Illus. 867. Test the fit of the hinge on a piece of scrap before drilling the work.

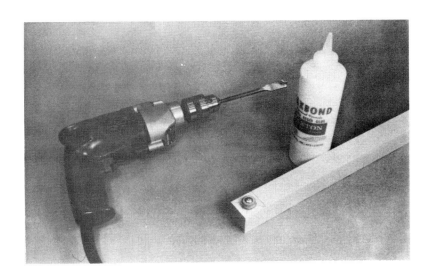

Illus. 868. Apply glue sparingly. Too much glue can damage the hinge.

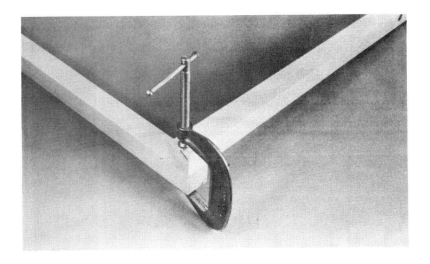

Illus. 869. Fit the parts together and clamp them securely. Do not remove the clamps until the glue has cured.

Work Platforms Work platforms or supports are used between sawhorses (Illus. 870) to support stock or saws such as the Compound Cut Saw or motorized mitre box (Illus. 871). The size of the work platforms depends on the intended use. Generally, they are 16–24 inches wide and 6–8 feet long. Keep them small enough to be light; this will make transporting them easier.

Study the drawing (Illus. 872) to determine how the parts fit together. After you have decided on overall dimensions, you can develop a bill of materials. Crosscut the framing members to length and assemble them with glue and nails. Check the assembly to be sure that it is square.

Cut the plywood top slightly oversize. Attach it to the framing members with glue and nails. Trim the plywood to size with a laminate-trimming router bit or by belt-sanding the edges. Do this after the glue cures. Apply a coat of penetrating oil to all surfaces; this will help reduce the chance of warping and will protect the plywood top from the elements.

Another form of work platform or work support is the Shop System™ table; this table is used with the saw-guiding system described in Chapter 6 (Illus. 873 and 874). This table is built in the same fashion except that the framing stock and the top are somewhat heavier. The ends have no supports; this is where the Shop System clamps to the base. Study the drawing (Illus. 875) and read the manufacturer's instructions before beginning.

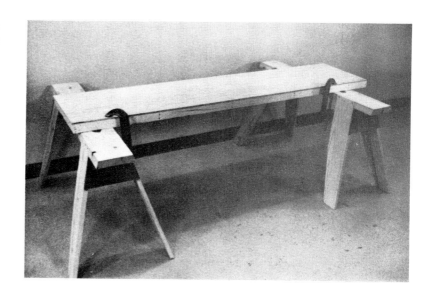

Illus. 870. When cutting, fitting, or machining stock, put a work support between sawhorses to support the stock.

Illus. 871. The work support also provides a clamping surface for mitring devices.

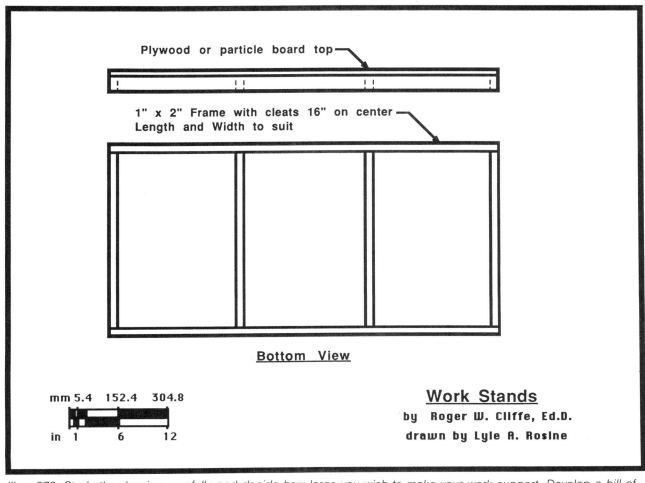

Illus. 872. Study the drawing carefully and decide how large you wish to make your work support. Develop a bill of materials from your overall dimensions.

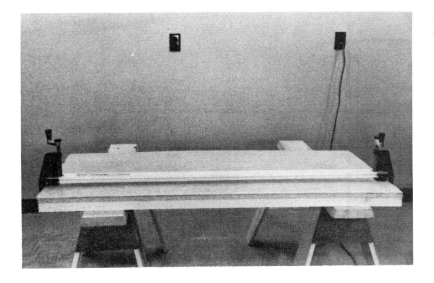

Illus. 873. This work support is used with the Shop System.

Illus. 874. The Shop System controls the path of the saw. Clamp it to the work platform.

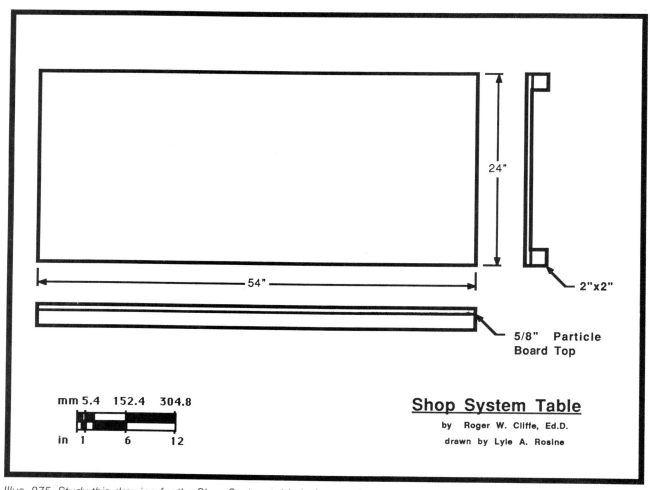

Illus. 875. Study this drawing for the Shop System table before you begin building it. It would be wise to order the Shop System before building the platform.

Siding Jig or Mitre Box The siding jig or mitre box is a device that controls the path of the saw and provides a firm support for the work (Illus. 876). When in use, it is supported by sawhorses (Illus. 877). The saw rides in a track or support (Illus. 880).

The size of the siding jig or mitre box depends on the type of work you are doing. You may want to make the jig smaller than the one shown in Illus. 878, 879, and 881. Adjust the size to your needs. The jig must also be adjusted to the saw you will be using in the jig. It is best to check these things before beginning.

Cut the support members and plywood base. Rip and rabbet the front and back pieces. Lay out the tapers and saw them. Smooth the exposed edges of the front and back. Glue and nail the front and back to the base. Keep the front and back perpendicular to the base. Check this with a square.

While the glue is curing, make the saw supports. Study Illus. 881 and select some straight-grained hardwood. Cut and notch the saw supports. Fasten them to the front and back with flathead screws. Make sure that they are perpendicular to the front and back. They should also be spaced correctly for the saw you intend to use. If you wish, you could use angle iron instead of wood for the saw supports or guides.

To make the siding jig a mitre box, install a set of supports or guides at a 45-degree angle. This will allow you to cut face and some end mitres.

Protect the jig with one or two coats of penetrating oil. Handle it carefully to maintain its accuracy.

Illus. 876. This jig can be used for crosscutting or mitring. Siding installers frequently use a jig like this.

Illus. 877. Support the jig with sawhorses when using it. The sawhorses position it at a working height.

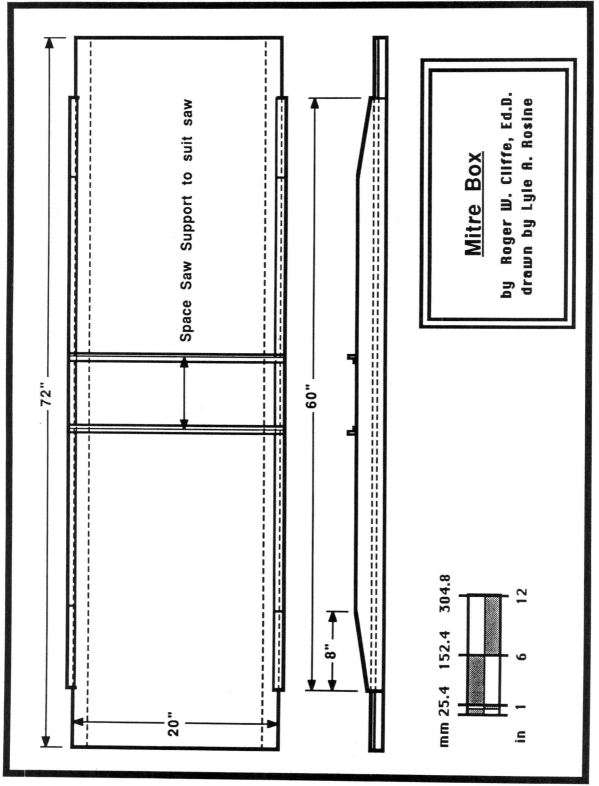

Illus. 878. This drawing gives the overall dimensions of the siding jig or mitre box. Use the drawing here and in Illus. 879 and 881 to develop a bill of materials. The saw supports can be shifted to one end. This allows for a second set of supports at a 45-degree angle.

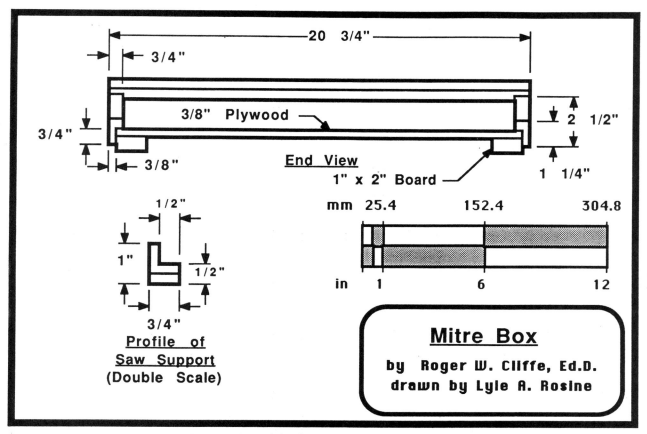

Illus. 879. This drawing gives the side view of the jig.

Illus. 880. The jig is designed to control your portable circular saw on rails. The saw's path is guided by the rails. The rails can be set at any desired angle.

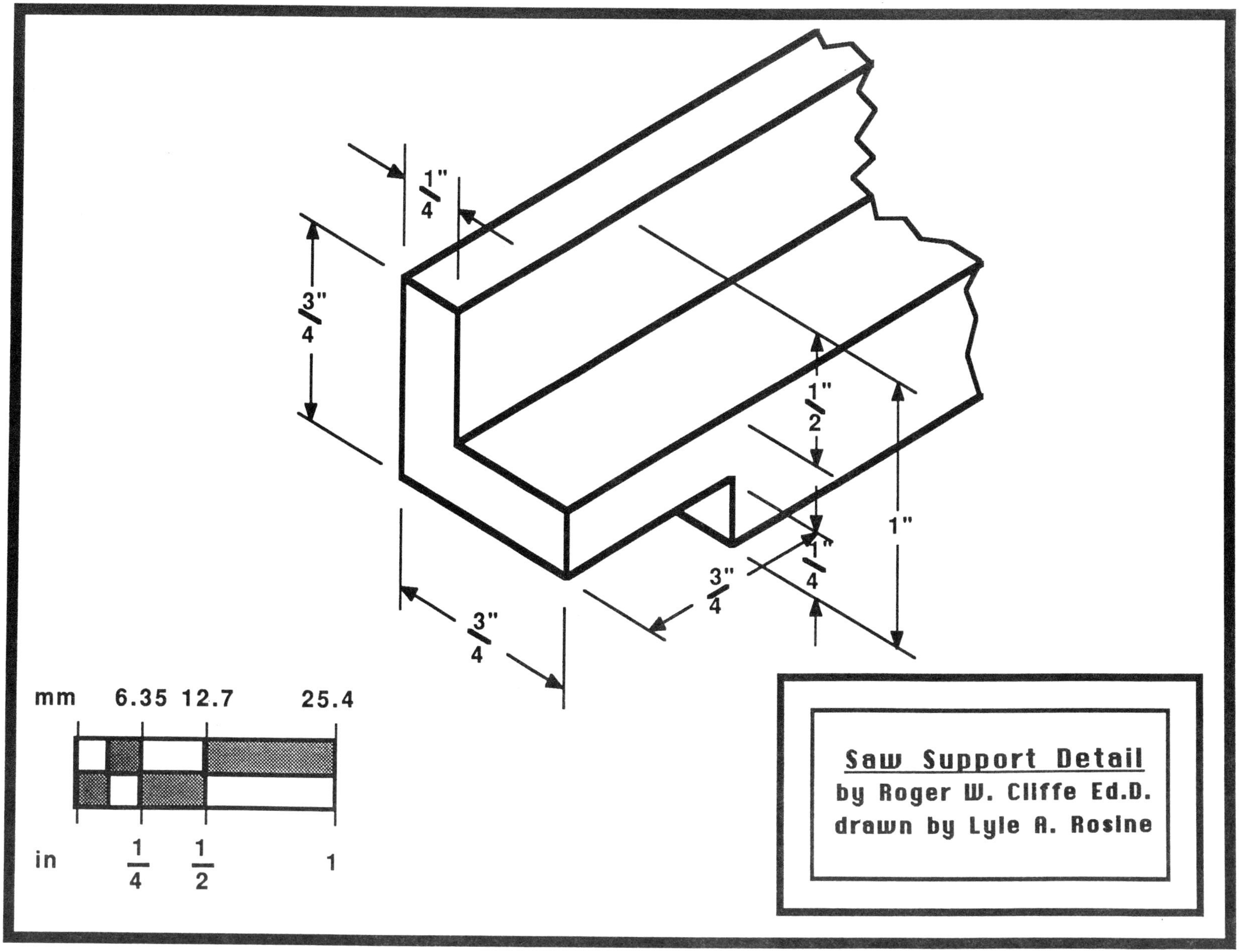

Illus. 881. This drawing gives you a detail of the saw supports. Use a straight-grained hardwood for these supports.

Saw Case A saw case is a desirable accessory that you can build (Illus. 882). The case can both protect your saw and store extra blades (Illus. 883). You can store wrenches, lubricants, or other accessories in compartments inside the case (Illus. 884). Begin by studying the plans (Illus. 885–887). The drawings suggest a direct-drive saw, but all dimensions are letters instead of sizes. You provide the dimensions based on your saw. If you want to see a worm-drive case, refer to Illus. 139–141.

Illus. 882. The saw case makes a desirable accessory for a portable circular saw. A custom-built saw case is sure to meet your design requirements.

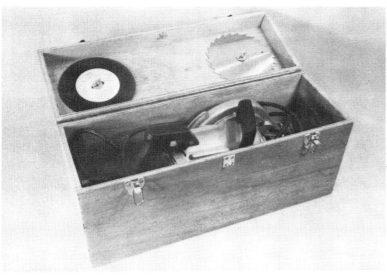

Illus. 883. Extra blades can be stored under the lid. Wing nuts hold the blades securely in position.

Illus. 884. You can store lubrication, wrenches, and other accessories in special compartments in this case.

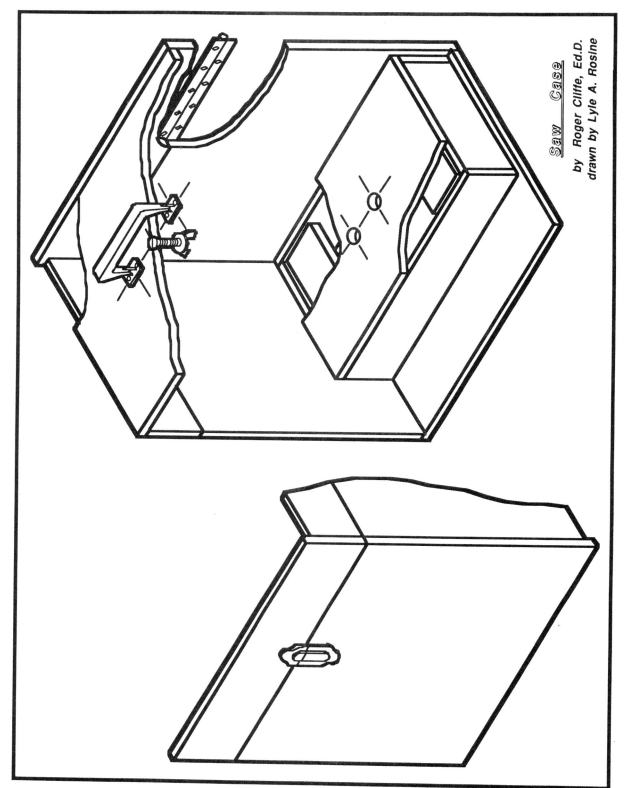

Illus. 885. This assembly drawing shows how the parts fit together. Study this drawing before you develop your bill of materials.

Saw Case
by **Roger Cliffe, Ed.D.**
drawn by **Lyle A. Rosine**

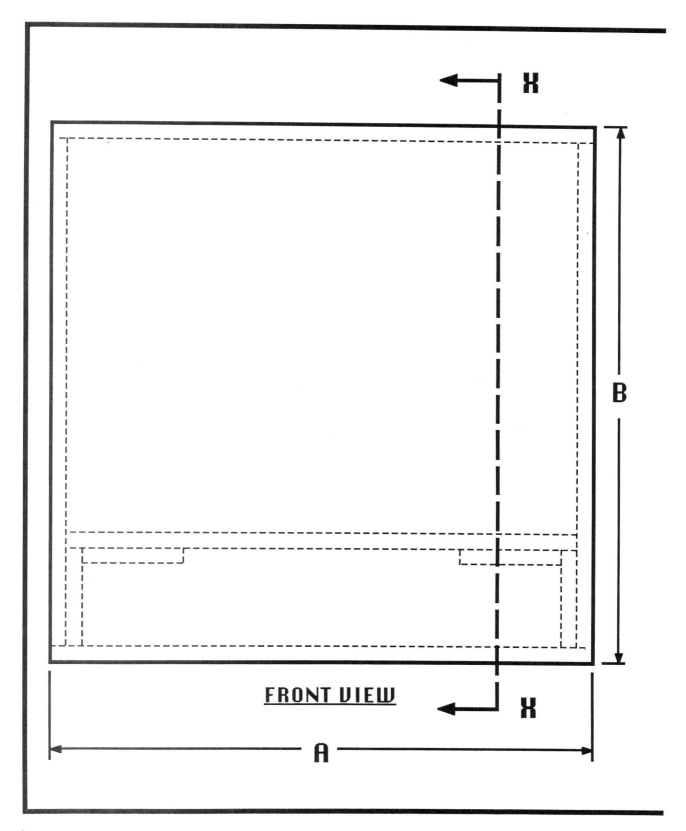

Illus. 886. The front view shows how the parts are related. Replace the letters with dimensions that fit your saw.

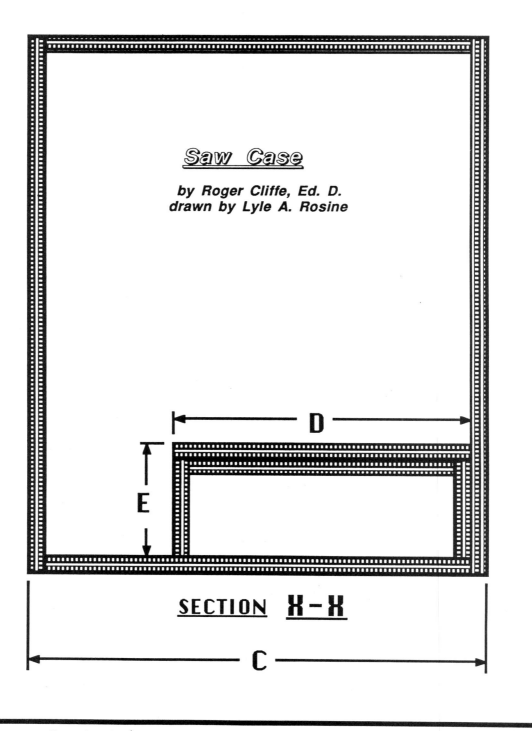

Illus. 887. Incorporate dimensions in this section view to suit your saw. Use these dimensions to develop your bill of materials.

The box that supports the saw in the section view is high enough to clear the guard when the blade is fully exposed. The width of the support allows the base of the saw to ride on it fully without the motor hitting the side of the case.

After you have determined the dimensions of your saw case, develop a bill of materials. Use high-quality plywood to build your case. Be sure to make the sides a little wider than necessary; this way you can saw the box open after assembly.

Glue and nail the box together. Install the interior parts after you have sawn the box open. Sand all the parts (Illus. 888) after the glue cures. All plywood edges should be smooth.

Lay out the cutting line and clamp the box to a sawhorse or other work support. Tack a cutting guide to the box (Illus. 889) and cut across the end (Illus. 890). Do not cut through the plywood (Illus. 891). Cut both ends, and then cut the long sides (Illus. 892). Cut completely through the box on the long cuts (Illus. 893). The lid will remain intact (Illus. 894) because some stock remains on the ends.

Use a hand saw to cut through the ends (Illus. 895). True-up the ends where the hand saw opened the box (Illus. 896). Now fit the interior parts. The lids on the boxes will just sit on the opening. Fasten a cleat to the underside of the lids; this keeps them from sliding around. Drill two holes in the lids so that they can be lifted out easily.

If you expect to use your saw case roughly, you may want to add wooden fenders (Illus. 897); they will reinforce the box and absorb shocks. The brass corners also keep the wood from direct contact with the ground or moisture. They are a nice addition.

Hinge the box with a piano hinge. Two draw catches help pull the lid down securely. Lay out the catches carefully before drilling any holes. Install a padlock hasp for security, if desired.

If you wish to store blades under the lid, attach flathead screws to the top with T-shaped nuts. A wing nut can now be used to secure the blades to the lid.

A penetrating oil finish works well on a saw case. It can be touched up easily when the finish becomes scratched or dented. Two coats should be adequate.

Illus. 888. After the glue cures, sand the box carefully. All plywood edges should be smooth.

Illus. 889. Clamp the saw case to a sawhorse and tack a saw guide to the layout line.

Illus. 890. Cut across the end first. Set the blade depth before making the cut.

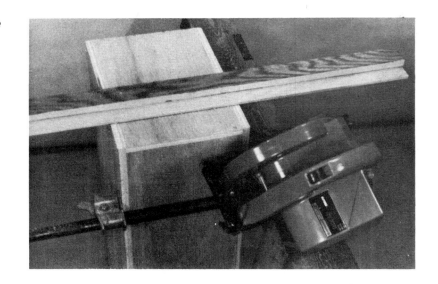

Illus. 891. The cut should not go through the side of the saw case.

Illus. 892. Cut through the long sides. Blade depth should be adjusted through the sides.

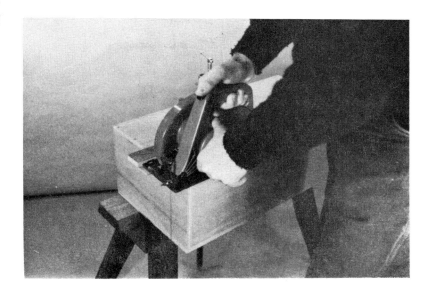

Illus. 893. Hold the saw carefully. Make sure that it follows the saw guide. An accurate cut is very important to the fit of the lid.

Illus. 894. The lid will remain intact after you have cut both long sides. A small amount of wood holds the lid at the ends.

Illus. 895. Saw the box open with a hand saw. The ends will have some roughness where the saw opens the box.

Illus. 896. Trim the ends with a plane or file. Check the fit of the lid to the box to ensure a perfect fit.

Illus. 897. You can glue and nail wooden fenders to the box for extra strength. The decorative metal corners also hold the box up off the ground.

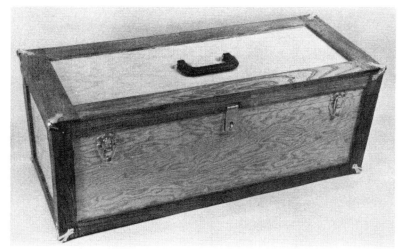

Mitre-Box Case The mitre-box case is a good project for the trim carpenter (Illus. 898). It has wheels and handles for portability. You can bolt the saw to the top or lock it inside the case, which means that it can be left overnight on the job. A drawer can be added to hold things like a coping saw, rasp, nails, nailset, and other related accessories.

Study the drawing (Illus. 901) and develop a bill of materials. The box shown here was built from ⅝-inch particle board. The case was designed for a 15-inch mitre box. If you have a 10-inch box, you want to make the case smaller.

After you have cut all the parts, glue and screw the box together. Use particle board or drywall (Illus. 900) screws. Bolt the wheels to the bottom of the case (Illus. 902). Lay out the holes carefully before drilling. Glue and clamp a support under the top near the front. Add screws after the part has been clamped in the correct position.

Bolt the doors to the case using strap hinges. A barrel bolt will hold the left door in position. Then mount the padlock hasp between the doors. Lay out the position of the mitre box on the top and inside of the case. Drill the holes through and install T-type nuts. Then bolt the mitre box to the case.

If desired, you can build a drawer now (Illus. 903). You can also mount a clamp bolt to the case; this allows stock to be clamped while you are cutting a cope joint (Illus. 904).

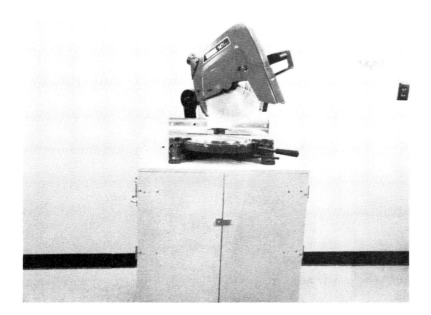

Illus. 898. This mitre box case is a good project for the trim carpenter.

Illus. 899. The motorized mitre box can be locked in the case when it is not being used.

Illus. 900. Drywall screws join particleboard parts together well.

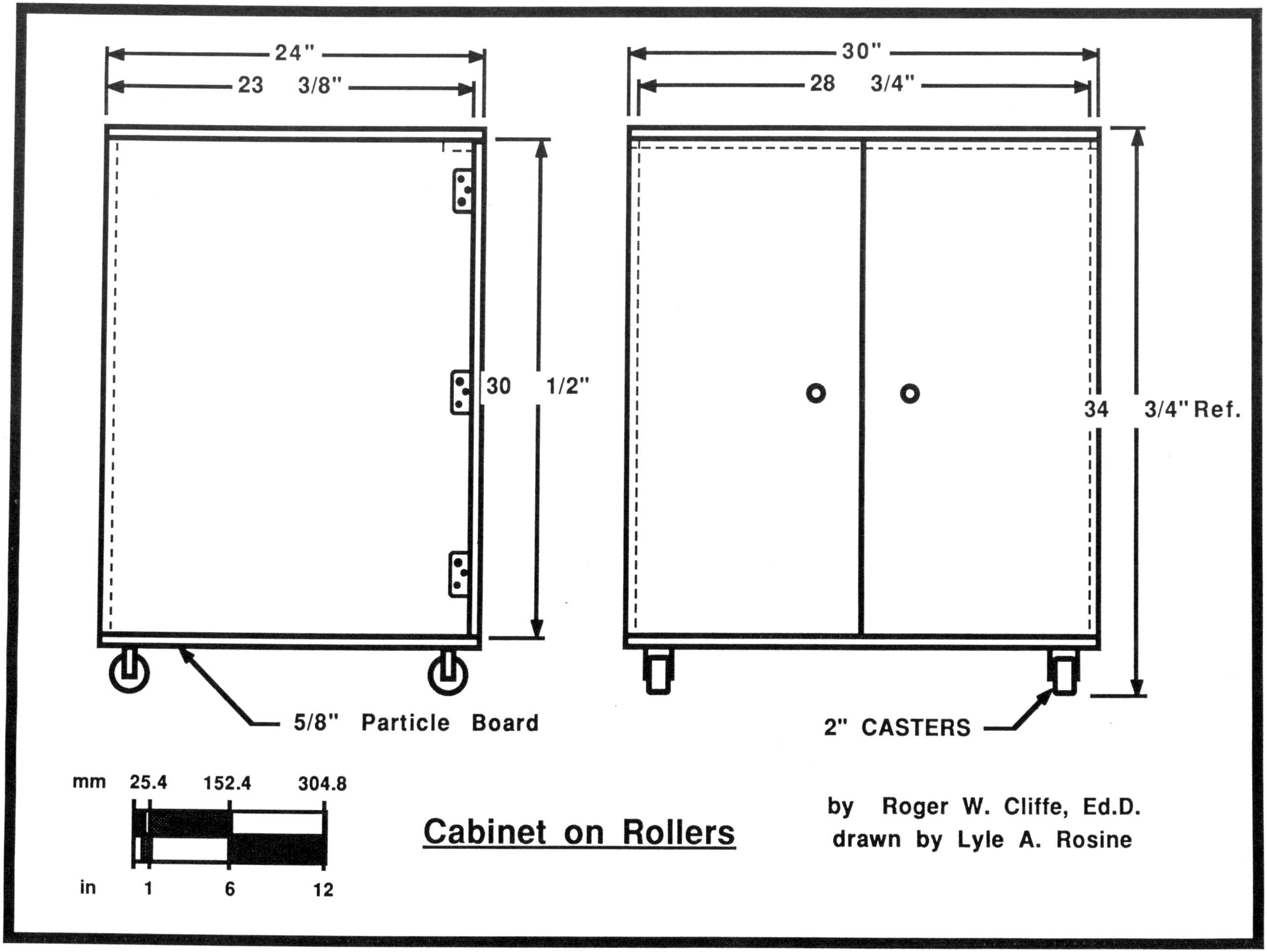

Illus. 901. Study the drawing. Make any needed modifications and develop the bill of materials.

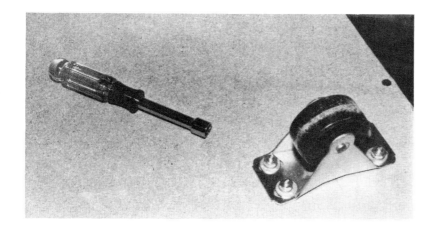

Illus. 902. Bolt the wheels to the bottom of the case to make it portable.

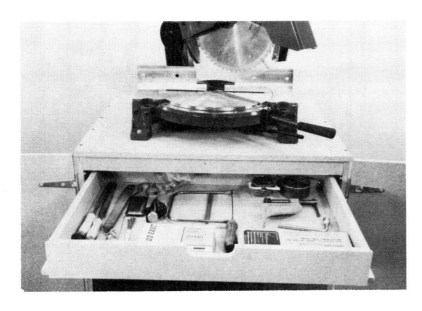

Illus. 903. If you want, you can add a drawer for storage. The drawer holds the necessary trimming accessories.

Illus. 904. You can add a clamping device to the case. This helps secure the stock when you are cutting cope joints.

Portable Workbench The portable workbench (Illus. 905) was designed to fold up for compact storage and easy transportation to the job. The clamping device holds stock securely for sanding or other operations. When it is not in use, it can be stored in the tool trough (Illus. 906).

Study the cutaway drawing (Illus. 909). This will help you see how the parts fit together. Decide how large you want to make the table, and develop a bill of materials. Make the top from sheet stock, the borders from hardwood, and the cleats that stiffen the table and anchor the brackets from framing material.

It is best to buy all the hardware before you begin; this way, you can make allowances for the position of the folding legs, the vise, and the hold-down clamp.

Begin by cutting the top to size. Build up one edge and screw the bottom of the tool trough to the spacers (Illus. 907). Then glue and nail the ends in position. Position the stiffeners and glue them in position; nails or screws can be used for reinforcement. Locate the legs on the stiffeners and screw them (Illus. 908) in position. Roundhead wood screws work best for this task (Illus. 910).

Decide on the position of the hold-down bolts, and drill the appropriate holes. First, counterbore a hole for the head of the bolt, and then drill a pilot hole for the bolt (Illus. 911). Counterbore the stiffener to accommodate the nuts and washer (Illus. 912), and adjust the bolt correctly (Illus. 913).

Chisel away the stock (if necessary) for correct fit of the vise (Illus. 914). Attach the vise with wood screws (Illus. 915) and line the jaws with a hardwood such as birch or maple (Illus. 916). This protects the work when it is secured in the vise (Illus. 917).

Sand the bench as needed. Radius the sharp corners and apply a finish. Use an oil finish (Illus. 918) that does not chip off or discolor the work. The oil also resists the adhesion of any glue that may be spilled.

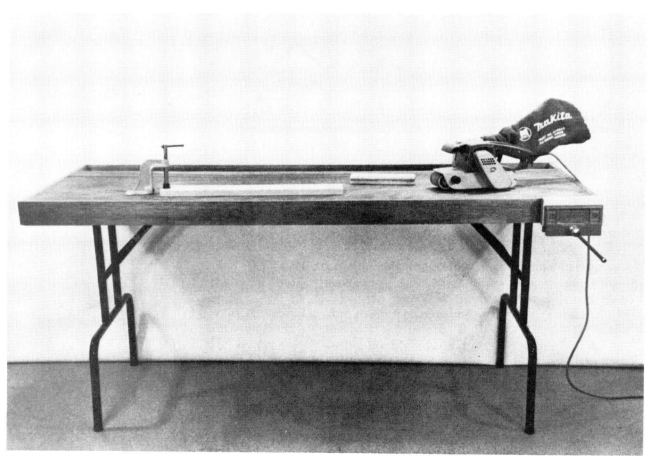

Illus. 905. This workbench folds up for easy storage and transportation. It also has a hold-down clamp and vise for holding stock.

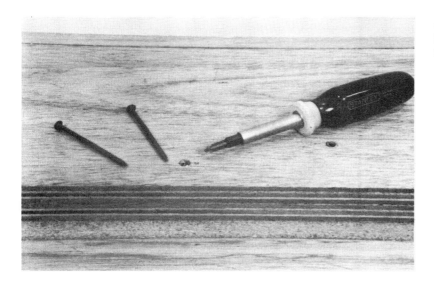

Illus. 906. You can store tools in this tool trough; it also keeps the wood surface clear.

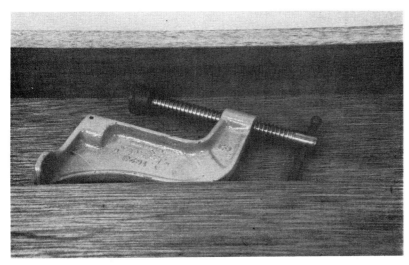

Illus. 907. Screw the bottom of the tool trough to the spacers. The tool trough is about 2¼" deep.

Illus. 908. Locate the legs on the stiffeners and drill pilot holes.

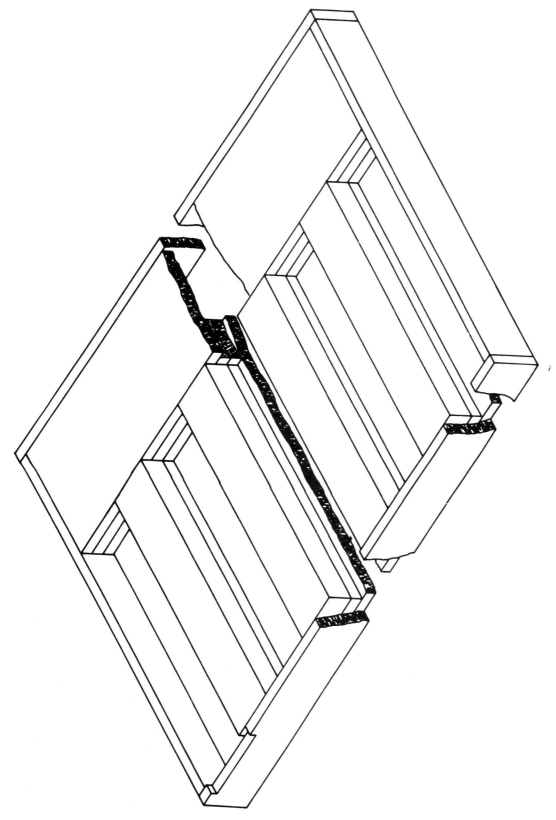

Illus. 909. This cutaway drawing shows how the parts fit together. The hardware you purchase and the size of the bench you want will determine your bill of materials.

Illus. 910. Round-head wood screws work well for securing the legs to the stiffeners. Sheet-metal screws can also be used.

Illus. 911. Drill a pilot hole in the middle of the counterbored hole. The counterbored hole holds the bolt head when the clamp is not in use.

Illus. 912. Counterbore the stiffener to accommodate the washer. The pilot hole drilled from above keeps all the holes aligned.

Illus. 913. Adjust the bolts so that the bolt head rises the correct distance above the workbench.

Illus. 914. Chisel stock away so that the vise fits correctly. You may have to add additional blocking to reinforce the vise.

Illus. 915. Use lag bolts to hold the bottom of the vise in position, and wood screws to hold the front.

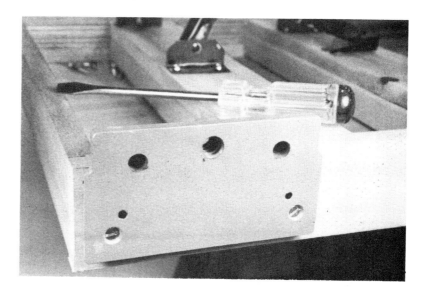

Illus. 916. Line the vise jaws with hardwood to protect the work.

Illus. 917. Make sure the hardwood liners line up. Also, make sure that the jaws extend beyond the end of the bench. This allows clearance for sawing.

Illus. 918. An oil finish works well on the workbench. It is easy to repair, does not chip, and glue does not stick to it.

Appendices

METRIC EQUIVALENCY CHART

MM—MILLIMETRES   CM—CENTIMETRES

INCHES TO MILLIMETRES AND CENTIMETRES

INCHES	MM	CM	INCHES	CM	INCHES	CM
1/8	3	0.3	9	22.9	30	76.2
1/4	6	0.6	10	25.4	31	78.7
3/8	10	1.0	11	27.9	32	81.3
1/2	13	1.3	12	30.5	33	83.8
5/8	16	1.6	13	33.0	34	86.4
3/4	19	1.9	14	35.6	35	88.9
7/8	22	2.2	15	38.1	36	91.4
1	25	2.5	16	40.6	37	94.0
1 1/4	32	3.2	17	43.2	38	96.5
1 1/2	38	3.8	18	45.7	39	99.1
1 3/4	44	4.4	19	48.3	40	101.6
2	51	5.1	20	50.8	41	104.1
2 1/2	64	6.4	21	53.3	42	106.7
3	76	7.6	22	55.9	43	109.2
3 1/2	89	8.9	23	58.4	44	111.8
4	102	10.2	24	61.0	45	114.3
4 1/2	114	11.4	25	63.5	46	116.8
5	127	12.7	26	66.0	47	119.4
6	152	15.2	27	68.6	48	121.9
7	178	17.8	28	71.1	49	124.5
8	203	20.3	29	73.7	50	127.0

YARDS TO METRES

YARDS	METRES	YARDS	METRES	YARDS	METRES	YARDS	METRES	YARDS	METRES
1/8	0.11	2 1/8	1.94	4 1/8	3.77	6 1/8	5.60	8 1/8	7.43
1/4	0.23	2 1/4	2.06	4 1/4	3.89	6 1/4	5.72	8 1/4	7.54
3/8	0.34	2 3/8	2.17	4 3/8	4.00	6 3/8	5.83	8 3/8	7.66
1/2	0.46	2 1/2	2.29	4 1/2	4.11	6 1/2	5.94	8 1/2	7.77
5/8	0.57	2 5/8	2.40	4 5/8	4.23	6 5/8	6.06	8 5/8	7.89
3/4	0.69	2 3/4	2.51	4 3/4	4.34	6 3/4	6.17	8 3/4	8.00
7/8	0.80	2 7/8	2.63	4 7/8	4.46	6 7/8	6.29	8 7/8	8.12
1	0.91	3	2.74	5	4.57	7	6.40	9	8.23
1 1/8	1.03	3 1/8	2.86	5 1/8	4.69	7 1/8	6.52	9 1/8	8.34
1 1/4	1.14	3 1/4	2.97	5 1/4	4.80	7 1/4	6.63	9 1/4	8.46
1 3/8	1.26	3 3/8	3.09	5 3/8	4.91	7 3/8	6.74	9 3/8	8.57
1 1/2	1.37	3 1/2	3.20	5 1/2	5.03	7 1/2	6.86	9 1/2	8.69
1 5/8	1.49	3 5/8	3.31	5 5/8	5.14	7 5/8	6.97	9 5/8	8.80
1 3/4	1.60	3 3/4	3.43	5 3/4	5:26	7 3/4	7.09	9 3/4	8.92
1 7/8	1.71	3 7/8	3.54	5 7/8	5.37	7 7/8	7.20	9 7/8	9.03
2	1.83	4	3.66	6	5.49	8	7.32	10	9.14

Glossary

Actual, or Rated, Horsepower. The saw's power under load when the blade is cutting.

Arbor Hole. The hole in the saw blade that is used to mount it on the machine. Though most arbor holes are circular, some have a diamond shape.

Base Lubricant. A self-adhering tape-like material that is applied to the base of the portable circular saw. It is designed to prevent scratching.

Blades. See individual blades.

Carbide-Tipped Blades. Blades with teeth made from small pieces of carbide. They are much harder and more brittle than the steel used for conventional blades. They are also more expensive, but require much less maintenance. Carbide-tipped blades come in the following classifications: rip, crosscut, combination, hollow-ground, and plywood.

Chain-Saw Blade. Specialty blade that is a combination circular-saw and chain-saw blade. Its kerf is about twice as wide as that made by a conventional circular-saw blade.

Chop-Stroke Mitring Machine. Mitring machine in which the blade comes down on the work.

Clutch-Drive Portable Circular Saw. Portable Circular Saw that has a round arbor and a mating round hole in the blade. A spring washer is used between the outer arbor washer and the arbor nut. This washer allows the arbor to slip if the blade becomes jammed or pinched in the work.

Combination Blade. A blade designed for both ripping and crosscutting that cuts mitre joints very well. Combination blades come in two different designs: Some have teeth that come to a point and a rip-tooth profile, and others have a chisel edge and a smaller hook angle.

Compound Mitre. A mitre joint with two angular cuts.

Compound-Mitre Mitring Machine. A chop-stroke mitring machine that allows for blade tilt. It can cut compound mitres in a flat plane.

Control-Cut Blade. Carbide blade with 8–12 teeth that are set slightly above the blade's periphery. These teeth minimize the chance of a kickback or severe cut.

Control Devices. Devices like straightedges, clamps, and push sticks that make sawing easier and safer. When you use control devices, not only are your hands well away from the cutting area, you also have control of the job.

Cope Joint. A joint used in many jobs to fit inside corner moulding and door stops. A cope joint looks like a mitre joint, but, unlike the mitre joint, wood shrinkage does not cause the cope joint to open up.

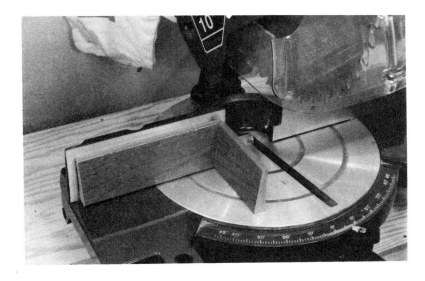

This cope joint fits well, yet it requires no hand sawing.

Corner Lap Joint. A lap joint that has two identical parts. It looks like an elongated rabbet.

Crosscut. A cut made across the grain of the workpiece.

Crosscut Blade. A blade that cuts across the grain. Crosscut blades have smaller teeth than rip blades; these teeth come to a point, not an edge.

Cross-Lap Joint. A lap joint with two identical parts that look like dadoes.

Dado Joint. U-shaped channel made with or across the grain.

Developed Horsepower. The saw's power when the blade is turning, but not cutting.

Diamond Blades. Metal discs with diamond particles bonded to their sides. These discs can be either solid or segmented. Some are used for specific materials such as glass.

Direct-Drive (Helical-Drive) Portable Circular Saw. Portable Circular Saw in which the motor is perpendicular to the blade. Direct-drive saws have an arbor that extends directly from the end of the motor or through the reduction gear. The blade is attached directly to the arbor.

Direct-drive saws usually have an arbor that extends from the motor. The blade is usually perpendicular to the motor.

Dovetail Lap Joint. An intermediate lap joint with a dovetail configuration.

Dust Collector. An adapter that fits over the sawdust chute and connects to the common vacuum hose. It cleans up the dust made by the portable circular saw.

Edge Mitre. A rip cut with the blade set to the desired mitre angle.

End Lap Joint. See Corner Lap Joint.

End Mitre. A mitre cut usually made at a 90-degree angle to the edge of the work.

Expansion Slots. Slots found between the teeth on most blades. These slots allow the blade to expand as it heats up when it is being pushed into the work, and keep it from warping.

Face Mitre. A mitre cut made across the face of the grain. It is similar to a crosscut.

Fence. An attachment to the portable circular saw that is used to control the saw during rip cuts. It usually clamps to the base or shoe and slides in a track for adjustment purposes. In the case of stationary circular machines, the fence actually clamps to the table.

Groove Joints. Dado cut made with the grain.

Ground Fault Circuit Interrupter. Used in many field operations to reduce electrical danger when a circular saw is being used, the GFCI measures the amount of current going to the tool and returning to the GFCI. When there is an imbalance in the current going and returning, a ground fault exists. A ground fault interrupts the current going through the GFCI.

Gullet. The area behind the cutting edge of the tooth. It carries away the sawdust cut by the tooth.

Heeling. Occurs when the blade does not go through the workpiece squarely. Heeling can cause rough cuts and tear-out.

Hollow-Ground Blades (Planer Blades). Blades with no set. The sides of the blade are recessed for clearance in the kerf. Hollow-ground blades should be used to cut mitres and compound mitres, and *not* used for heavy ripping.

Hook Angle. The angle of the tooth's cutting edge as it relates to the center line of the blade. Blades that are used for tough-cutting jobs sometimes have *negative* hook angles. Some circular-saw blades designed to cut used lumber have a negative hook angle so that they can cut nails or other metal in wood.

Hopper Cut. See Compound Mitre.

Intermediate Lap Joint. A lap joint with one part that looks like a rabbet and one that looks like a dado.

Joints. See individual joints.

Kerf. The cut made by a circular-saw blade. The kerf must be slightly larger than the saw-blade thickness.

Kerf Splitter. A saw helper that keeps the kerf open while the saw is cutting.

Lap Joints. Are classified as either corner or intermediate lap joints. A corner lap joint has two identical parts. The cut looks like an elongated rabbet. Intermediate lap joints have one part that looks like a rabbet and one that looks like a dado.

Masonry-Cutting Discs. Discs, usually made from silicon carbide, used to cut masonry.

Metal-Cutting Discs. Discs, usually made from aluminum oxide, used to cut metal.

Mitre Box. See Siding Jig.

Mitre Cuts. Angular cuts made across the face, end, or edge of the work. Most mitre cuts are made at 45-degree angles so that when the two pieces are joined they make a 90-degree angle.

Mitre Table- and Radial-Arm-Saw Adapters. A device that transforms a portable circular saw into a table saw or other stationary cutting guide.

OSHA (Occupational Safety and Health Act). Regulates the setup and use of industrial equipment. This act is enforced by the Occupational Safety and Health Administration, which is part of the United States Government. OSHA lists specific requirements for portable circular saws. They should be observed faithfully.

Parallel-Base (Drop-Shoe) Portable Circular Saw. A portable circular saw with a base that slides on machine ways near the front of the saw. The parallel-base saw keeps the motor parallel to the base as you reduce blade exposure.

Pitch. A brown, sticky substance that accumulates on a circular-saw blade when it becomes hot. As pitch accumulates on the blade, it prevents it from dissipating heat and causes it to become dull more quickly.

Begin a plunge cut with the lower guard retracted. Once the blade has come up to full speed, slowly lower it into the work.

Plunge-Cut. A blind cut or rectangular hole made with the portable circular saw that is used on electrical outlets, sink cutouts, and plumbing and heating access holes.

Plywood Blades (Panelling or Veneer Blades). Blades with very fine crosscut teeth with little set that are designed to cut hardwood plywood with cabinet- or furniture-grade outer veneers.

Positive-Drive Portable Circular Saw. A saw in which the engagement of a diamond-shaped arbor with a blade with a mating diamond-shaped arbor eliminates slipping of the blade on the arbor.

Pull-Stroke Mitring Machine. A mitring machine that has a motor that travels on a pair of metal rods; it is pulled into the workpiece when it cuts. Pull-stroke mitring machines are usually compound-mitring machines.

Rabbet Joint. An L-shaped channel made along the edge of the work.

Rip Blade. A blade with a straight-cutting edge that is designed to cut with the grain. Rip blades have deep gullets and large hook angles.

Rip Cut. A cut made with the grain.

Heavy rips are best done with a worm-drive saw or a heavy-duty direct-drive saw. Note that the work has been clamped to the sawhorse.

Saw Case. An accessory that helps keep the saw dry, clean, and free of corrosion.

Saw Guides. Devices used to guide the path of the portable circular saw. They help the operator to cut a straight line.

Siding Jig. A device that controls the path of the saw and provides firm support for the work.

The single or simple-mitre machine has no provision for blade tilt.

Single-Mitre Mitring Machine. A chop-stroke mitring machine that has no provision for blade tilt. It can cut compound mitres if the stock is tilted.

Splitter. A piece of steel used during rip cuts that attaches to the blade saw behind the blade. The splitter follows the blade into the wood during the rip cut and keeps the saw kerf open as the cut progresses.

Taper Cut. A common portable circular saw cut made on furniture legs, as well as braces and wedges.

Tilt-Base (Pivot-Shoe) Portable Circular Saw. A portable circular saw that has a point at the front or back of the saw on which the blade pivots. The balance of the blade changes when you change the blade exposure.

Tooth Set (or Offset). The bend in the blade's teeth that allows the blade to cut a kerf that is larger than the blade's thickness.

Top Clearance. Downwards slope of the back of the blade tooth. This slope keeps the back of the tooth from rubbing on the wood. Without top clearance, the blade cannot cut.

Triangular Blade. A triangular-shaped specialty blade made from tool steel that can make irregular or curved cuts.

Work Supports. Devices on which the work is positioned for cutting. Sawhorses and saw benches are two common types.

Worm-Drive Portable Circular Saw. Portable circular saw in which the motor is parallel to the blade. Worm-drive saws have a gear mechanism between the motor and the blade.

The motor on a worm-drive saw is parallel to the blade. Worm-drive saws have a positive drive, so the blade does not slip or spin freely.

Index

A

Accessories. *See also specific accessories*
   attachments, 56–60
   compatibility, 24
   helpers, 60–68
   projects, 278–331
Accessory tables
   for mitre boxes, 204–209
   for pull-stroke mitring machine, 252–254
Accidents, factors contributing to, 20–21
Actual horsepower, 46
Adapters, 21
Adjustment of saw
   general considerations, 157–158
   safety and, 24
Advanced and specialized operations. *See specific operations*
Alignment, of chop-stroke mitring machine, 212
Ampere ratings, 46
Angles, odd, determining mitre cuts for, 202–203
Anti-kickback clutch, 10–11
Arbor bearings, lubrication, 158
Arbor hole, 29
Arbor knockout, 29
Arbor speed, 48
Arcs, cutting, 105
Arm, on mitring machine, 168
Attachments, 56–60. *See also specific attachments*

B

Balance of saw, 162
Base
   evaluation, 162
   size, 46, 48
   types of, 54
   warpage, 154
Base lubricant, 59–60
Basic operations. *See specific operations*
Batteries, charging, 49
Battery-powered portable circular saws, 11, 12, 49
Bearing type, 162
Belt changing, for chop-stroke mitring machine, 214
Bevel angle, 239
Bevel post settings, type of cut and, 189
Binding of saw in kerf, 72, 76, 77
Biscuit joiner, 256–258
Blades
   for battery-operated saws, 49
   changing, 155–157
      for chop-stroke mitring machines, 209–210
      pull-stroke mitring machines, 243, 246–248
   maintenance, 41–44
   safe working distance from, 25
   selection, 40–41
      for laminated sheet stock, 97

sharpening, 43–44
size, 13
terminology, 27–30
types of, 30–40
Blind cut, 100–102
Bowed pieces, 176
Brick-mould corners, 194–195
Brushes, changing, 159
Bushings, 29–30
Buying a saw. *See* Purchasing of saw

C

Carbide-tipped blades
  characteristics and applications, 35–39
  dull, 42
  evaluation of, 38–40
Casey Hand Tool, 10, 256–257
Ceramic cutting, 130–133
Chain-saw blade, 34, 35
Chisel-tooth combination blade, 32
Chop-stroke mitring machines
  accessories and features, 16, 168, 170–172
  alignment, 212
  applications, 164
  blade changing, 209–210
  blade selection, 40–41
  brick-mould corners, 194–195
  characteristics, 164
  classifications, 164–167
  compound mitre cutting, 185–189
  cope joint making, 189, 192–194
  crosscutting with, 176–180
  crown moulding, cutting of, 189, 190
  general trim installation, 194, 199–202
  maintenance, 209–214
    belt changing, 214
    electrical, 213–214
    heeling, 213–214
    of table, 211
  mitre-box picture frames, 194, 196–198
  parts, 167–169
  purchasing, 214–215
  safety guidelines, 173–176
  simple mitre cutting, 176, 181–184
  stop mechanism and clamping mechanisms, 166–167
Circles, cutting of, 105

Clamping mechanism
  for mitring machine, 168, 170–171
  for pull-stroke mitring machines, 221–223, 226–228
Classification of saws, 50–56
Clothing, 24
Clutch drive, 55
Combination blades, 30–31, 32
Combination chop-stroke and push-stroke mitring machine, 258–259
Compound-angle settings for popular structures, 239
Compound-cut saw, 17, 41, 217, 219
Compound mitres
  cutting, 87, 89–90
    with chop-stroke mitring machine, 185–189
    with pull-stroke mitring machine, 238–245
  uses for, 112
Compound-mitre saws
  features, 13, 164–165, 168
  simple mitre cutting, 181–182, 184
Compressed-air saws, 49–50
Concrete cutting, 127
Control-cut blades, 37
Control devices, 25–26
Cope joint making, using chop-stroke mitring machine, 189, 192–194
Cordless portable circular saws, 11, 12, 49
Cords, replacement of, 159–160
Cracked saws, 26
Crain Model 800 Super Saw, 16, 17, 18
Credo blade, 37
Crosscut blades, 27, 30, 31, 32
Crosscutter, 257–259
Crosscutting
  with chop-stroke mitring machines, 176–180
  general procedure, 70–75
  with portable stationary saw, 265, 269–270
  with pull-stroke mitring machines, 224–230
Crown mouldings
  cutting method for chop-stroke saws, 189, 190
  cutting method for compound cut saws, 241–243
Cuda blade, 38
Cutting and finishing jig, 140–144
Cutting blades. *See* Guides

D

Dado joint, 107–108
Delta Sawbuck. *See* Sawbuck
Developed horsepower, 46
Development of circular saw, 10–19
DeWalt Crosscutter, 257–259
Diamond blades, 35
Direct-drive saws, 50, 52, 55
Doors, trimming, 120, 125–126
Double-insulated saws, 11
Dovetail lap joint, 112–115
Drop-shoe design, 14
Dull blades, 32, 41–43
Dust collection devices, 56, 59, 173

E

Edge mitres, 86–87, 112
Electrical cords
    extension, 21, 46
    inspection, 154
    replacement of, 159–160
Electrical maintenance
    for chop-stroke mitring machines, 213–214
    general, 159–160
    for pull-stroke mitring machines, 251–253
Electrical safety procedures, 21–22
Electrical shock
    compressed-air portable saws and, 50
    safety devices for, 11, 12
Electric brakes, 11
Electrical power, 48–49
End mitres
    cutting with pull-stroke mitring machine, 235–238
    general cutting procedure, 82, 85
Expansion slots, 30
Extension cords, 21, 46
Extension tables
    for compound-cut saw, 230
    for mitring machine, 168, 170, 176

F

Face mitres
    cutting with pull-stroke mitring machine, 231–234
    general cutting procedure, 82–84

Fatigue, accidents and, 20
Fences, 56–58, 219
Flip Saw, 261–265
Flipstop, 205, 209
Flooring blade, 32
Framing members, cutting notches in, 116–124
Framing/rip combination blade, 32
Friction blade, 32

G

Glass cutting, 130–133
Grinding-type cutters, 127
Ground Fault Circuit Interrupters (GFCI), 21–22
Ground-fault protection, 25
Guards
    for chop-stroke mitring machines, 170, 172
    inspection, 155
    for pull-stroke mitring machines, 219, 220
Guides
    for crosscutting, 72, 74–75
    general characteristics, 60, 63–67
    procedure for making, 278–280
    setting up, 98
    for stationary saws, 133–140
Gullet, 27, 28

H

Handles
    evaluation, for purchasing of saw, 162
    for mitring machines, 172–173
    types, 52–53
Hardwood plywood, cutting of, 97
Heeling
    adjustment for, 249, 251
    description, 213–214
Helical-gear saws, 50, 52. *See also* Direct-drive saws
Helpers, 60–68. *See also specific helpers*
History of circular saw, 10–19
Hollow-ground blades, 31–33
Hook angle, 27–28
    negative, 28–29
Hopper angles, 241
Horsepower, actual and developed, 46
Housekeeping, general, 21

I

Inattention to job, 20
Inca radial arm saw, 270, 273–274
Irregular shapes, cutting, 102–104

J

Jig
    cutting and finishing, 140–144
    siding, 144–145
    Workmate cutting, 144, 146–148
Joinery cutting, with portable circular saw
    compound mitre, 112
    dado, 107–108
    lap, 108–111
    mitre, 112
    rabbet, 105–107

K

Kerf, 27, 28
KerfKeeper, 60, 63
Kerf splitters, 60, 63, 76
Kickback, avoidance of, 55, 79

L

Laminated sheet stock, cutting of, 97
Lap joint, 108–115
Laser blade, 38, 39
Laser X2 blade, 37–38
Left-hand saws, 53, 54
Lion Measuring Bench, 194, 197
Lubrication
    general procedure, 158–159
    of pull-stroke mitring machines, 252

M

Maintenance
    blade, 41–44
    for chop-stroke mitring machines, 209–214
    general care, 154–161
    for pull-stroke mitring machines, 243, 246–252
Masking tape, to prevent tear-out, 97
Masonry cutting
    procedure, 130–132
    safety guidelines, 127
Metal- and masonry-cutting discs, 34–35
Metal cutting
    general procedure, 127–130
    using plunge-type mitring machines, 204–205
Metal-cutting blade, 32
Michel Electric Handsaw Company, 10
Microprocessor-controlled portable circular saw, 11
Miter Maker, 133, 134–136
Mitermate cutting guide, 75
Mitre angle, 239
Mitre-box case, shop-make, 322–325
Mitre boxes
    accessory tables for, 204–209
    making of, 311–314
    motorized, 13, 15, 16
Mitre-box picture frames, 194, 196–198
Mitre cuts
    determining for odd angles, 202–203
    procedure, 82–87, 243
        for edge mitres, 86–87
        for end mitres, 82, 85
        for face mitres, 82–84
    simple, with chop-stroke mitring machine, 176, 181–184
Mitre joints, 112
Mitre saw adapters, 65
Mitre saws, positioning of, 23
Mitring machine, specialty, 215–216
Motorized mitre box, 13, 15, 16
Motors, 11

N

Negative hook angle, 28–29
Ni-cad battery, 49
Nonwooden materials, cutting, 127–133
Norsaw, 261
Notches, cutting in framing members, 116–124

O

Offset, 27
Operating control, 26
OSHA regulations, 25–26

P

Parallel base, 54, 55
Particleboard/plywood blade, 37
Pigtail, 21
Piranha blade, 37, 38
Pitch, 41
Pitch angle, 239
Pitch removers, 41, 42
Pivot base, 54
Pivot-foot design, 14, 15
Pivot mechanism, for pull-stroke mitring machines, 221–223
Pivot shoe, 54. *See also* Tilt base
Planer blade, 31–33
Plastic cutting, using plunge-type mitring machines, 204–205
Plug, inspection and maintenance, 154–155
Plunge-cutting, 100–102
Plunge-cutting saws
  characteristics, 55–56
  crosscutting with, 72–73
  use of, 101–102
Plunge or chop-stroke mitring machines, 164–215
Plunge-type mitring machines, metal and plastic cutting, 204–205
Plywood, cutting of, 90–97
Plywood blades, 32, 33
Plywood-sawing jig, making, 304–307
Pneumatic saws, 49–50
Portable powered tools, OSHA regulations, 25–26
Portable stationary saws
  radial arm saws, 270–276
  table saws, 265–271
Portable workbench, shop-make, 326–331
Positive drive or clutch drive, 55
Power
  adequacy, determination of, 161
  types of, 48–50
Projects
  mitre box, 311–314
  mitre-box case, 322–325
  plywood-sawing jig, 304–307
  portable workbench, 326–331
  saw bench, 295–303
  saw case, 315–322
  saw guides, 278–280
  sawhorses, 278–294
  siding jig or mitre box, 311–314
  tips, 278
  work platforms, 308–310
Pull-stroke mitring machines
  adjustments, 246, 249–251
  applications, 217
  blade changing, 243, 246–248
  common operations, 224–252
  compound mitre cutting, 238–245
  crosscutting, 224–230
  Delta Sawbuck, 217–218
  electrical maintenance, 251–253
  end mitre cutting, 235–238
  face mitre cutting, 231–234
  lubrication, 252
  maintenance, 243, 246–252
  parts, 219–223
  safety guidelines, 221, 224
  types, 217
Purchasing of saw
  chop-stroke mitring machine, 214–215
  evaluation criteria, 160–162
  general considerations, 160–162
  selection criteria, 160
Push-pull saw, 259–261

R

Rabbet joint, 105–107
Rack system, 204–205
Radial arm saws
  adapters for, 65
  devices for, 153
  portable, 19
  portable stationary, 270–276
Rectangular hole cut. *See* Plunge-cutting
Right-hand saws, 53
Rip blades
  characteristics, 27, 30–32
  dull, 42
Ripping
  avoiding kickbacks, 79
  heavy, 77, 79–82
  procedure, 75–82
  of small pieces, 77, 78
  using portable stationary saw, 265, 269
Rpm's, 161, 164

Ryobi radial arm saw, 275–276

S

Safety
electrical, 21–22
evaluation of, 160–161
guidelines
for chop-stroke mitring machines, 173–176
for cutting nonwooden materials, 127
general, 24–25, 69
Saw bench, making, 295–303
Sawbuck, 17, 217–218, 219, 251
Saw case, 60–62
making, 315–322
Saw guides. *See* Guides
Sawhelper Miter-Grid, 194, 198
Sawhelper Ultrafence, 205, 207–209
Sawhorses, procedure for making, 278–294
Saw-Mite, 153
Saw Shop System, 133, 137–140
Saw size, 46–48
Saw Slik, 60
Scarf joint, 184–185
Sharpening of blades, 43–44
Sheet stock
cutting of, 90–96
laminated, cutting of, 97
Shoe, 54. *See also* Base
Shoe lubricant, 59–60
Shop-made accessories, 209
Shop System table, 308–310
Siding jig
description, 144–145
procedure for making, 311–314
Silicon carbide disc, 34–35
Simple-mitring machines, 164, 181–184
Single-mitre machine, 164, 181–184
Size of saw, 13, 162
Skil Corporation, 10, 11
Skilsaw, 10
Slate cutting, 127, 130–132
Specialty circular saws, 256–259
Splice joint, 184–185
Splitter, 56, 59
Squaring of blade, 249, 250
Stationary saws

circular, portable and unique design, 259–265
devices for, 140–153
guides for, 133–140
Stock, inspection of, 24
Stone cutting, procedures, 130–132
Stop, 228–229
Straight cuts, 98–99
Super Saw, 16, 17, 18
Switches, 26
replacement of, 159–160

T

Tables
accessory
for mitre boxes, 204–209
for pull-stroke mitring machine, 252–254
maintenance, for chop-stroke mitring machines, 211
for mitring machine, 164, 165, 167
for pull-stroke mitring machines, 219–221
Table-saw devices, 144, 149–152
Table saws
adapters for, 65
portable, 19
portable stationary, 265–271
Taper cuts, 87, 88
Tear-out, 90, 94, 96–98
Thorness blade, 33–34
Thrust handle, 52–53
Tilt base, 54
Tilt-base saw, 54
Tool-steel blades, 90
Tooth set, 27, 28
Top clearance, 29
Top handle, 52–53
Triangular blade, 33–34
Trim
base or ceiling, splice or scarf joint for, 184–185
crown moulding
cutting procedure for chop-stroke saws, 189, 190
cutting procedure for compound-cut saws, 241–243
general installation, 194, 199–202
Trimming doors, 120, 125–126

U

Underlayment, cutting plywood and sheet stock for, 90–96

V

Versatile Saw, 261

W

Warpage, 154

Warped or twisted stock, crosscutting of, 227
Wedges, 76
Weight of saw, 46–47, 162
Working environment, general, 22–23
Workmate cutting jig, 144, 146–148
Work platforms, making, 308–310
Work supports, 65, 67–68, 90–91, 93, 230
Worm-drive saws
   general characteristics, 50–53
   lubrication, 159
   pivot base, 54
   positive-drive system, 55

PHOTO CREDITS

The illustrations, tables, and charts in this book display the products, creations, and photography of many people and business organizations. Represented among them are the following: **Adjustable Clamp Company**, Illus. 143, 144; **AEG Power Tool Corporation**, Table 1, Illus. 1, 111 and 112; **American Design and Engineering**, Illus. 498, 602, and 728; **Black and Decker U.S. Power Tools Group**, Illus. 14–16, 20–22, 77–79, 108; **Casey Manufacturing Co.**, Illus. 2, 3, 731; **Crain Cutter Company**, Illus. 25–28; **Delta International Machine Corporation**, Illus. 24, 488, 491, 497, 621, 622; **Foley Belsaw Company**, Illus. 45–50, 52, 53, 60, 62, 63, 75, 88, 89; **Forrest Manufacturing**, Illus. 85, 91–93; **Hirsh Company**, Illus. 158; **Hitachi Power Tools USA, Ltd.**, Illus. 106; **Makita Electric Tools, USA**, Illus. 17; **Mafell North America, Inc.**, Illus. 9, 30, 742; **Matrix Enterprises**, Illus. 178, 205, 553; **Milwaukee Electric Power Tools**, Illus. 13, 43A, 242; **O Mark Industries**, Illus. 76, 80, 81; **Norsaw**, Illus. 743; **OMGA Construction Machinery**, Illus. 617–619; **The Pippin Corporation**, Illus. 437–439; **Porter-Cable Professional Power Tools**, Illus. 5, 12, 114, 116, 136; **Pootatuck Corporation**, Illus. 567; **Power Tool Institute**, Illus. 35, 36, 44; **Rockwell International Industrial Air Tools**, Illus. 107; **Sears, Roebuck and Company**, Illus. 41, 43B, 61, 110, 149, 259, 261, 373–375, 431–434, 454, 456, 541, 625, 651, 657–659, 685, 716, 718, 719; **Skil Corporation**, Illus. 1, 6, 31, 113, 260, 455; and **Welliver and Sons**, Illus. 397, 399, 400, and 403

Acknowledgments

Portable Circular Sawing Machine Techniques represents the work and cooperation of many people. A project of this magnitude is seldom done by the author alone. Photography and custom darkroom work were done by Bill Peter's Photography. Special thanks to photo models Steve Holley and Chris Sullens.

The drawings for jigs and projects were done by Mr. Lyle Rosine. His work makes project construction much easier for the reader.

Typing and revision work were handled by Loveda Paulus. Loveda: Many thanks for your help and commitment to deadlines. I hope you are enjoying retirement.

The patience and love extended by my wife, Cathy, and son, Austin, were greatly appreciated. Cathy and Austin: Thanks for your understanding.

Also greatly appreciated are the commercial photographs, line drawings, and products furnished by the following people and organizations:

Jeff Cowie, Adjustable Clamp Co.
James Frawley, AEG Power Tools
T. David Price, American Design and Engineering
John Padbury and Jim Roberts, Black and Decker U.S. Power Tools Group
Robert Blake, Carbide Saw Manufacturing, Laser Sawblade
Robert Jakubowski, Casey Manufacturing
Millard Crain, Crain Cutter Company
Gene Sliga, Delta International
Steve Eckard and Fred Garms, DML, Inc.
John Baenisch, Foley Belsaw Co.
John Potter and Jim Forrest, Forrest Manufacturing
Barry Dunsmore, Freud Blades and Tools
Conrad Blasko, Hirsh Company
Rich Nidaira, Hitachi Power Tools
Steve Friestedt, Makita Power Tools
Charlie Small, Matrix Enterprises
Jerry Wright, Milwaukee Power Tools
Michael Lovett, Omark Industries
G. Ghizzoni, OMGA Construction Machines
John Niswonger, Pippin Corporation
Philip Nathnagle, Pootatuck Corporation
Rick Carr and Rick Schmidt, Porter Cable
James E. Bates, Power Tool Institute
William Haselden and Jim Norberg, Rockwell International
Bill Peel and Steve Holley, Ryobi Power Tools
Mike Mangan, Sears, Roebuck and Company
Jim Maloney, Skil Power Tools
Dick Riggins, Wisconsin Knife Works